PONDS

AN ILLUSTRATED GUIDE

American Bullfrog
Lithobates catesbeianus

PONDS

AN ILLUSTRATED GUIDE

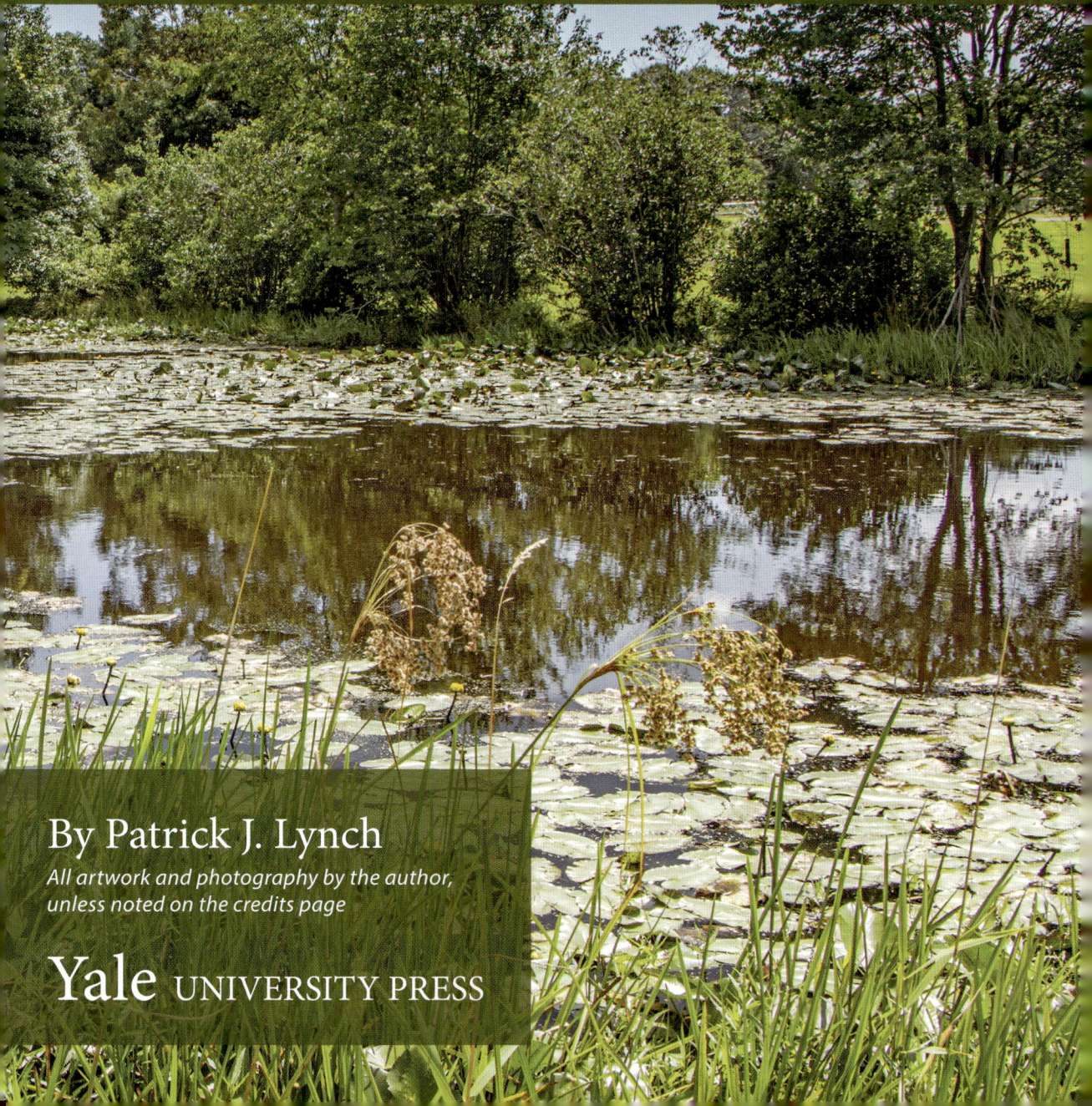

By Patrick J. Lynch

All artwork and photography by the author,
unless noted on the credits page

Yale UNIVERSITY PRESS

To Devorah, Zubin, and Arya.

To my wife, Susan Grajek, my mentor, muse, wise woman, and best friend.
I could not have done it without you.

Yale University Press books may be purchased in quantity for educational, business, or promotional use. For information, please e-mail sales.press@yale.edu (US office) or sales@yaleup.co.uk (UK office).

Printed in China.

ISBN 978-0-300-28496-6   Library of Congress Control Number: 2025941393

Authorized Representative in the EU: Easy Access System Europe, Mustamäe tee 50, 10621 Tallinn, Estonia, gpsr.requests@easproject.com

10 9 8 7 6 5 4 3 2 1

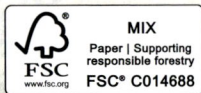

FSC
www.fsc.org

MIX
Paper | Supporting
responsible forestry
FSC® C014688

CONTENTS

Spring at Dismal Swamp State Park, South Mills, North Carolina.

PREFACE

This book is a general introduction to the natural history of ponds, small lakes, and other still freshwater environments in the Eastern United States from New England south to Florida. I have organized this book around environments, not particular locations. Although I focus on this region's freshwater plants, animals, and physical foundations, you cannot write about the natural world these days without constantly referencing the effects of environmental history and anthropogenic climate change. We live in the Anthropocene Epoch: human activity and climate change have become the dominant forces that shape our geophysical and biological environments.

This guide cannot be an exhaustive catalog of everything that lives in or near ponds and other freshwater environments—such a book would be neither practical as a guide nor very useful to the typical natural history reader, hiker, birder, kayaker, angler, or boater. I have emphasized the most dominant and common plants and animals, as well as the major biological principles at work in each environment. I intend to show you the major plants and animals that populate our freshwater environments so that you can walk into a wet meadow or along a pond edge and identify much of what you see. This is the first step in developing a deeper, more ecological understanding of our many and varied freshwater landscapes. This guide to the ***what*** of plants and animals also attempts to show ***why*** and how the human and natural aspects of the landscape coevolved into the freshwater wetlands we see today.

For recommended field guides to plants, wildflowers, geology, birding, insects, and other natural history topics, please see the works listed in "Further Reading."

Pat Lynch
North Haven, Connecticut

coastfieldguides.com
Bluesky: patrlynch1.bsky.social
Instagram: patrlynch1
Facebook: patrick.lynch1
patrlynch1@gmail.com

Spring at Brooksvale Park, Hamden, Connecticut.

ACKNOWLEDGMENTS

First and foremost, I thank Jean Thomson Black, Senior Executive Editor for Life Sciences, Physical Sciences, Environmental Sciences, and Medicine, for her suggestion to write this book. I mentioned to Jean that I wanted to shift away from regional field guides but that my previous five book projects had left me with hundreds of illustrations and tens of thousands of photographs. A large pile of potentially useful material, but what to do with it? Jean suggested that I turn my focus from regions (including Cape Cod, Long Island Sound, and the Connecticut River) toward a potential series of books focused on environments, such as ponds, streams, and beaches. Thus, the idea for this book was born. With a bit of luck, it might be the first of several.

I write books not because I know so much but because I love to learn, and I particularly love to create visual explanations of bioscience and natural history subjects. Over the years I have benefited enormously from my friendships with expert naturalists, distinguished scientists, fellow natural history illustrators, and friends who are eminent photographers. In particular, for decades I've profited from the wise media and computing advice of Rick Leone and the counsel on all things about natural history and birding of Connecticut's best birder, Frank Gallo.

David Skelly, the Frank R. Oastler Professor of Ecology, Director of the Peabody Museum of Natural History, and Professor of Ecology and Evolutionary Biology at the Yale School of the Environment, has lent wise advice on my last several books, and he has been generous with his time and counsel on this project, too. Mark A. McPeek, the David T. McLaughlin Distinguished Professor of Biological Sciences at Dartmouth's Department of Biological Sciences, also offered excellent advice on a number of topics in this book.

I particularly thank master field biologist and herpetologist Twan Leenders for once again lending me his magnificent frog, salamander, turtle, and snake photographs. These photo treasures add immeasurably to this book.

I also thank master wildlife photographer Judy Gallagher for her generosity in sharing her hundreds of superb photographs with the world via Flickr and Creative Commons licensing.

And I thank the manuscript editor on this project, Laura Jones Dooley, for her wisdom, expertise, and guidance on every page here. Laura is the best of colleagues: she's a genius whose advice makes me look smarter than I am.

I'm deeply grateful to these distinguished professionals, scientists, and my editors, but any mistakes here of fact or emphasis are my own.

Machimoodus State Park, Moodus, Connecticut.

Introduction to Ponds and Wetlands

Wetlands are some of our most important and biologically diverse environments. Characterized by saturated soils or standing water, these habitats support a remarkable array of plant and animal life while providing such essential services as water purification and flood control. Wetlands also play a critical role in limiting the growth of the atmospheric carbon compounds that cause global warming. In a process called carbon sequestration, wetlands bind up vast amounts of carbon in their vegetation, both in the living biomass of plants and in the large volume of dead and decaying plant material that underlies all wetlands.

Al's Pond, a coastal freshwater pond and marsh system just over 100 yards from Cape May's ocean beaches.

Freshwater wetlands in the Eastern United States can be broadly classified into four types: ponds and lakes, marshes, swamps, and bogs.

Ponds and lakes are standing bodies of water of various sizes, usually having a substantial portion of open water without emergent vegetation. In general, ponds are smaller and shallower than lakes. The light-filled, shallow water of ponds usually supports many rooted plant species (*see illustration, p. 6*).

Marshes are dominated by vegetation, such as cattails, reeds, grasses, and sedges. Marshes are typically found in areas with shallow standing water or saturated soils. Marshes with many shrubs are called shrub marshes or shrub swamps (*see "Freshwater Marshes"*).

Swamps are dominated by substantial trees and shrubs. A typical swamp consists of a wet forest floor or pool overshadowed by such large, water-tolerant tree species as Red Maple, Black Tupelo, and Yellow Birch. Thick tangles of shrubs including willows, dogwoods, alders, and Buttonbush dominate some swamps. Swamps are found in areas with poorly drained soils and often occur near ponds. They usually show abundant standing surface water in the wetter seasons of spring and fall. In the heat of summer, the swamp surface may look

Wetlands of all kinds are extremely valuable components of a healthy landscape because they provide so much varied and productive habitat for plants and animals. Here, a tiny but fascinating Round-Leaved Sundew grows in a rural bog area in central Massachusetts. The survival of such unusual varieties of plants is possible only when we preserve all forms of wetlands from filling, development, or destruction.

Hawley Bog Preserve, Hawley, Massachusetts.

like a dry forest floor, but the soil underneath the leaf litter remains wet (*see "Freshwater Swamps and Shrub Swamps"*).

Bogs are acidic wetlands that accumulate a thick layer of sphagnum moss (peat). They are typically found in cool, northern regions and are fed primarily by precipitation. Bogs support a unique community of acid-loving plants, such as sphagnum moss, sundews, cranberries, and pitcher plants. Because their waters are highly acidic, bogs rarely have substantial populations of aquatic animals, although the edges of bogs may be rich in wildlife (*see "Northern Bogs and Southern Pocosins"*). Many northern and high mountain ponds have boglike sphagnum communities around their edges (*see illustration, pp. 198–99*). These sphagnum mossy areas host many of the same unusual plant species found in true bogs.

The importance of wetlands

Wetlands provide critical breeding, nesting, and foraging habitat for a variety of wildlife, including birds, fish, amphibians, reptiles, mammals, and invertebrates. Permanent wetlands such as ponds and lakes serve as nurseries for many fish species and provide shelter and food for juvenile fish. Wetlands also act as natural water filters, removing pollutants, excess nutrients, and sediments from the water. The dense vegetation and microbial communities in wetlands break down pollutants and improve water quality. Wetlands are a critical component of flood control. They act as natural sponges, absorbing and storing excess water during heavy rainfall and snowmelt periods. The thick vegetation of marshes and swamps also acts to slow and absorb the force of floodwaters. The mechanical obstructions of plants and their roots reduce the risk of flooding downstream and protect property and infrastructure.

Marsh and swamp wetlands are important carbon sinks (areas that absorb and hold carbon), storing large amounts of carbon in their soils and vegetation. This stable, long-term storage of biological carbon is called carbon sequestration. Carbon storage mitigates climate change by removing carbon dioxide from the atmosphere. Both marine and freshwater wetlands are increasingly recognized as crucial elements in controlling the atmospheric carbon dioxide that causes most global warming.

Wetlands provide opportunities for various recreational activities, such as wildlife watching, fishing, kayaking, and hiking, so much so that it's hard to imagine a natural park landscape without ponds or streams.

What is a pond?

Most biologists distinguish ponds and lakes less by overall size than by the water's depth and the patterns of vegetation both surrounding and within the water. In general, lakes are still-surface bodies of water

Healthy ponds and other wetlands aren't just pleasant-to-view components of the contemporary landscape. Ponds provide crucial habitats and sustenance for thousands of kinds of plants and animals. In a rapidly changing and unpredictable climate, wetlands also perform critical ecosystem services, protecting the developed landscape from stormwater flooding.

Like many of today's parkland ponds, this pond is largely artificial. The former farm pond was enlarged and landscaped when the State of Connecticut bought the farm about 20 years ago and transformed it into Machimoodus State Park. The pond has very healthy and productively clear water, well protected from the usual excess nutrients that plague most suburban ponds and streams. This pond is fed by small seasonal streams that run off a nearby wild forested hillside, with few pollution sources in the stream drainage. All ponds should be so fortunate.

Along with Smooth Alder, Redosier Dogwood, and Black Willow, Buttonbush (pictured) is very common in shrub swamps and the wet edges of ponds throughout the Eastern United States.

deep enough (more than 15 feet) that rooted aquatic plants cannot grow near their centers. Some lakes are small but deep, with only a rim of aquatic vegetation near their shorelines. Ponds are shallower bodies of water with extensive edges of emergent aquatic plants. Ponds usually also have rich submerged aquatic vegetation growing across or almost across their centers. Marshes are very shallow bodies of water covered almost entirely by emergent vegetation, such as grasses and sedges. In later successional stages, aquatic shrubs such as Smooth Alder, Buttonbush, or Silky Dogwood may convert grassy marshes to shrub swamps (*see illustration, pp. 162–63*).

Ponds in the Eastern United States exhibit a wide range of physical characteristics shaped by their geological origins, climate, and surrounding landscape. Glacial kettle ponds, formed by giant chunks of ice left behind by retreating glaciers, are common in the north, while artificial ponds, including farm ponds, dammed streams, and small reservoirs, are scattered throughout the region. Beaver ponds are unique natural water bodies where beavers create wooden dams that impound water to create living spaces not only for beavers but for many other aquatic animals (*see illustration, pp. 146–47*).

Pond size varies considerably, from smaller pools that dry up seasonally to larger shallow-water bodies that can span several acres. Pond depth also varies, with shallow littoral zones (near-shore areas) grading into deeper profundal zones (open water). Ponds are usually shallow enough (less than 15 feet deep) that rooted aquatic plants such as water lilies can grow across the open water surface. Daily temperatures vary widely in shallow ponds, as the water is quickly heated by summer sun or cooled by fall and winter winds.

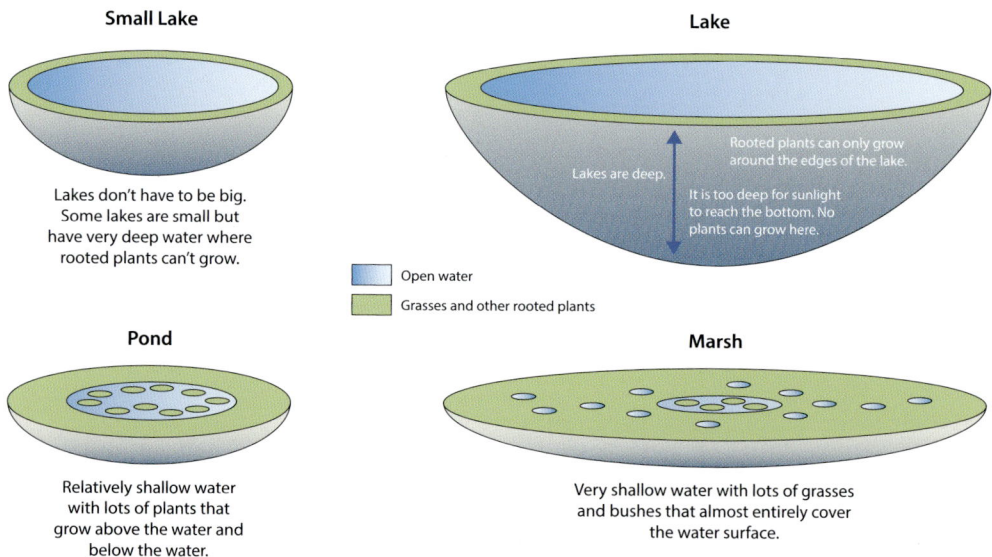

Small Lake

Lakes don't have to be big. Some lakes are small but have very deep water where rooted plants can't grow.

Lake

Lakes are deep.

Rooted plants can only grow around the edges of the lake.

It is too deep for sunlight to reach the bottom. No plants can grow here.

Open water

Grasses and other rooted plants

Pond

Relatively shallow water with lots of plants that grow above the water and below the water.

Marsh

Very shallow water with lots of grasses and bushes that almost entirely cover the water surface.

The origins of Eastern ponds

The retreat of glaciers at the end of the last Ice Age, formally called the Wisconsinan Glacial Episode, ended in the Northern United States and southern Canada approximately 12,000 years ago. Glacial ice was pivotal in forming freshwater ponds in the Northeast. As glaciers moved across the landscape, they sculpted the terrain, leaving behind depressions and basins in the earth. When the ice melted, these depressions filled with water, forming the ponds we see today. Some ponds originated as kettle holes, formed when ice blocks broke off from the main glacier, were buried by glacial drift, and eventually melted, leaving behind water-filled holes. The sandy southeastern coastal margins of New England are particularly rich in kettle ponds. The Cape Cod region alone has hundreds of kettle ponds.

In the unglaciated parts of the Eastern United States, natural ponds often originated in the slow but steady erosion of the Appalachian Mountains over the past 100 million years. Streams and rivers often formed natural dams that created lakes and ponds. Over millennia, these geological processes resulted in diverse freshwater ponds and lakes of varying size, depth, and water chemistry.

The North American Beaver also created countless small dams and ponds across the Eastern and Midwestern landscapes. Beavers once lived along most permanent streams and rivers, and in the past, their wooden dams transformed most streams and small rivers into chains of small beaver ponds. Beginning shortly after the European coloni-

Natural pond areas are almost always a mix of habitats. Here, a small pond also hosts a small but vibrant cattail marsh and a belt of swamp forest around the edges. This pond was created largely by the activity of beavers. The giveaway for the presence of beavers is the number of dead trees that emerge from the marshy area in the background. Beavers dammed the pond's feeder stream and created a small pond. The pond water saturated the soil under the trees, killing the trees. These dead trees are not wasted: they are valuable nesting and roosting sites for woodpeckers and other woodland birds as well as for classic pond birds such as the Wood Duck, which nests in tree cavities above water.

Most ponds in today's highly developed suburban landscape are artificial, as here in rural Lyme, Connecticut. In the eighteenth and nineteenth centuries, small stream dams were used to power grain mills. The mills are long gone, but the dams and ponds remain.

zation of North America, beavers were heavily hunted for their fur and were almost entirely extirpated from the Eastern United States. Once the beavers were gone, their dams disintegrated, and a massive acreage of ponds and wetlands was lost. Now that beavers are much less hunted, beaver ponds and marshes are reappearing in rural and forested areas.

In the modern, developed landscape of the Eastern United States, most ponds and lakes were created and actively shaped by human activity. Most larger lakes are human-created or modified with dams to act as regional water reservoirs. Before coal and steam power, small factories relied on water mill power to grind grain and power manufacturing machines. On most smaller rivers and streams, there are still thousands of dams in place that originated in the water-powered years of the 1700s and 1800s. In Connecticut alone, there are over 4,800 stream dams. Most dams along rivers created small ponds behind the mill dam. Virtually all these small dams have long been obsolete. Many are in poor repair and are an increasing flood hazard as storms have become more frequent and wetter due to climate change. Such small dams can suddenly breach under the load of a heavy rainstorm.

The small mill dams also prohibit anadromous breeding in formerly abundant species such as Alewife, Blue-Backed Herring, American Shad, and Atlantic Salmon.

THE FRESHWATER CYCLE

Freshwater can seem endlessly abundant on a fair early summer day along the East Coast. But on a global scale, the kinds of visible surface ponds, lakes, streams, and rivers we may take for granted are shockingly rare. Only 3.5 percent of the Earth's total water supply is freshwater, and most of that is bound up in the polar ice caps or groundwater. Ponds, lakes, rivers, and streams account for about 0.26 percent of global water supplies (*see illustration, below*).

The water cycle describes the large water storage zones in the environment and how water constantly moves among these zones. The atmosphere, surface waters, and groundwater are freshwater storage zones. Gravity, solar heating, wind, local weather, and the activities of plants are the major motive forces in shifting freshwater from one storage zone to another. Within the local and regional water cycles, water can regularly change among liquid, solid (ice), and gaseous (water vapor) forms (*see illustration, overleaf*).

Human activities modify both the quantity and quality of environmental water cycles. We dam major rivers such as the Connecticut, the Hudson, and the Susquehanna to create reservoirs that supply

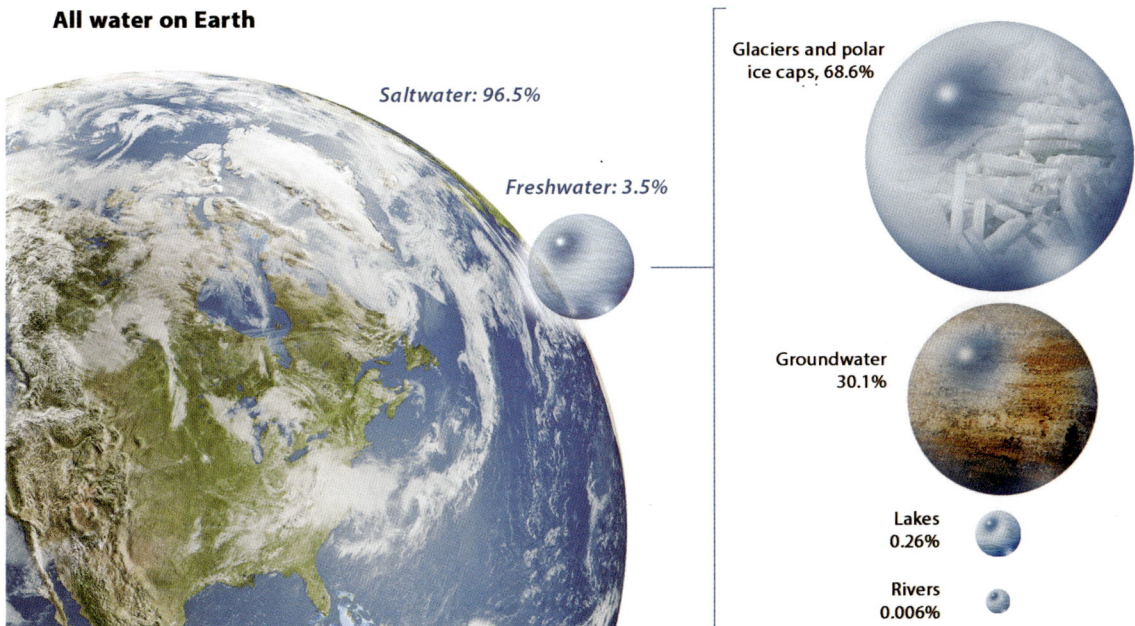

All water on Earth

Saltwater: 96.5%

Freshwater: 3.5%

Glaciers and polar ice caps, 68.6%

Groundwater 30.1%

Lakes 0.26%

Rivers 0.006%

Solar energy
Energy from the sun is the ultimate power source for the wind and the water cycle

THE FRESHWATER CYCLE

Evaporation is driven by solar heating of the land

Evaporation cycle

99% of rainfall returns to the atmosphere immediately as evaporation and transpiration from plants

Evaporation from land surfaces

Transpiration from vegetation

Runoff from land

Groundwater flow

Local water tab

Bedrock

Deep percolation of groundwater into bedrock layers

Wind, powered by solar energy and driven by passing weather systems, drives the atmospheric movements of water. Wind also transports atmospheric water to and from distant oceans.

Wind

Cloud formation and condensation

Precipitation cycle

Evaporation from lakes and rivers

Rainfall

Too many impervious surfaces, including roads, parking lots, building roofs, and concrete stormwater drains, can interfere with groundwater flows. This can contribute to lowering the local water table, and risks severe flooding in major storms because impervious surfaces and concrete drains flood with fast-moving water in heavy rain.

Groundwater outflow into lakes and rivers

our homes and businesses with water and electricity and to prevent flooding during major storms. Our domestic and industrial activities constantly introduce contaminants, pesticides, and fertilizers into the water cycle. Stormwater runoff from developed areas and agriculture carries sediment and pollutants into local waterways, where the contaminants eventually flow into regional river systems.

Transpiration and groundwater absorption are critical but normally invisible water cycle elements. As plants respire and use light and water to create photosynthesis, they "exhale" as much as 99 percent of the water they absorb through their roots. Tiny pores on the underside of leaves called stomata release water vapor into the air, where the water then returns to the atmosphere. Except in rare circumstances, streams, rivers, and lakes are fed mainly by groundwater that flows under the influence of gravity. In a typical rainstorm, only a small percentage of water reaches streams and ponds. Plant roots and soil absorb rainwater directly, and within hours, almost all of the rainwater has transpired back into the atmosphere or been absorbed into the groundwater. In relatively moist environments such as New England and the Mid-Atlantic Coast, groundwater reservoirs are often vast. Deep beneath the ground surface, the groundwater moves slowly through rock fissures and glacial soils. Depending on how porous the soils and

The Great Vermont Flood of 2023 was a destructive flash flood event that occurred in Vermont on July 10–11, 2023. The flooding was caused by a severe storm that dropped six to nine inches of rain over 48 hours, with the highest total, 9.2 inches, recorded in Calais. The storm affected all 14 counties and caused widespread damage, including two fatalities. Ironically, another rainstorm almost exactly a year later in July 2024 caused similar flood damage throughout central Vermont and New Hampshire. Climate change is likely to make such storms regular events in the Northeast.

bedrock are, groundwater can take tens or even hundreds of years to reemerge on the surface as it seeps into ponds, streams, and rivers.

The prominent exception to this normal rainwater cycle occurs during major rainstorms or sudden warming periods that melt the winter snowpack. If the soil and normal groundwater channels are saturated with water (or frozen in winter), rain or meltwater will run directly from the land into ponds, streams, and rivers. Usually, this direct runoff elevates ponds and stream levels temporarily. But in significant rainstorms, nor'easters, and hurricanes, so much rain running directly off the saturated ground and into local streams and rivers can cause catastrophic flooding. This is why the preservation of natural environments near streams and rivers is critical to flood control. Wetlands and forest soils act as giant sponges to absorb the effects of heavy rainfall. Roads, parking lots, storm sewers, concrete water drainage channels, and building roofs shed rainwater directly into local streams and rivers. The fast-flowing stormwater can turn ordinarily placid streams into a damaging flood within hours.

The Great Vermont Flood of July 2023 was a destructive flash flood event. Slow-moving thunderstorms produced heavy rainfall and flooding across the Northeastern United States, and the heaviest and most destructive flash flooding was centered around northern Vermont.

The regional freshwater cycles of transpiration, evaporation, condensation, and precipitation are only a tiny part of the larger continental and global water cycles. These larger water cycles are evolving quickly

Occasional severe droughts occurred in New England's past, even before recent changes in the climate. The New England drought of 1965 was a severe dry spell that affected the Northeast, particularly Massachusetts, Connecticut, and Rhode Island, during the summer. The drought was caused by a combination of factors, including a lack of precipitation, high temperatures, and low humidity.

Recently in various normally moist parts of the country, late-summer droughts have become routine. This is at least partly due to climate change and the warming environment, where the increased heat causes rapid evaporation and unpredictable rainfall patterns.

Drought in Massachusetts, 2000–2023

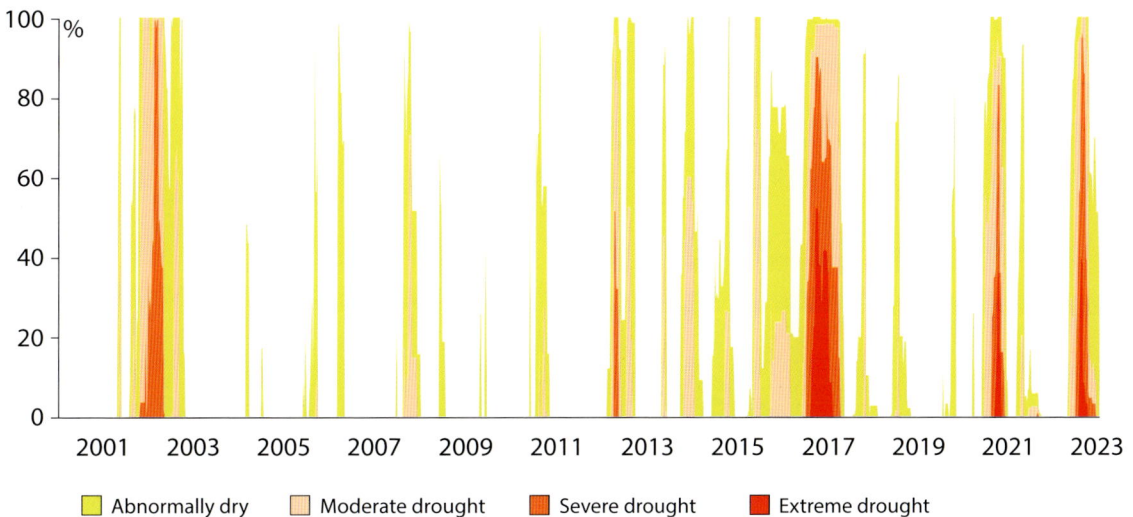

Abnormally dry Moderate drought Severe drought Extreme drought

U.S. Drought Monitor

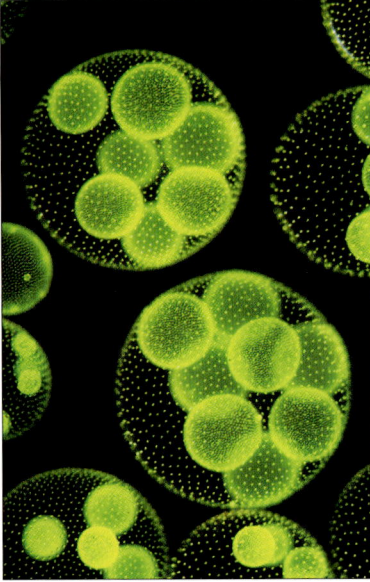

Microscopic green plants
(phytoplankton) are vital members of a healthy pond ecosystem. Green pond water is not necessarily an indication of pollution. But when ponds are polluted with fertilizer runoff from lawns and farms, excess populations of algae can cause problems. If pond water loses its transparency through excessive algae growth, the plants growing on the bottom of the pond can't support photosynthesis and may die.

because of climate change, and new global patterns of drought and flood emerge almost daily. Even historically moist Eastern North America isn't immune to climate change. In the late summer of 2022, much of New England and parts of the Mid-Atlantic and Southeast experienced a moderate to severe drought. Although rivers, lakes, and ponds didn't dry up and disappear in 2022, the weather pattern most adults grew up with changes in unpredictable ways every year.

THE LIFE CYCLES OF PONDS

Atmospheric oxygen readily enters the waters of ponds and streams, where it becomes the dissolved oxygen critical to aquatic life. This is particularly true in small, active streams, where the shade-cooled shallow waters and constant splashing and water movement help the waters absorb atmospheric oxygen. Even in the more stagnant waters of ponds and lakes, the wind drives small waves and water currents that bring fresh oxygen into the upper layers of the water column.

All plants and animals require oxygen for respiration and basic metabolic processes. In respiration, living cells use oxygen to process nutrient molecules and to support cellular energy cycles, growth, and maintenance. Living cells absorb oxygen and release carbon dioxide. In the sunlit part of the day, aquatic plants also produce oxygen as a byproduct of photosynthesis. This excess oxygen helps increase the level of dissolved oxygen in aquatic systems. However, plants also require oxygen for respiration, so during the night, the respiration needs of plants and aquatic animals reduce the dissolved oxygen level in ponds and streams.

The dissolved oxygen level that water can hold is tightly linked to water temperature—cold water can hold more dissolved oxygen than warm water. Both ponds and streams show a distinct seasonal cycle in

---

Daily photosynthesis and respiration cycle in a pond, late spring through midautumn

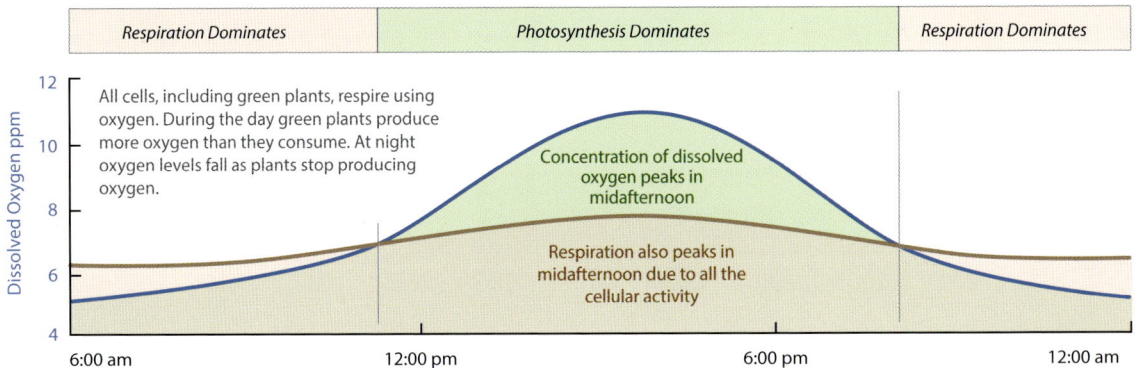

Respiration Dominates	Photosynthesis Dominates	Respiration Dominates

All cells, including green plants, respire using oxygen. During the day green plants produce more oxygen than they consume. At night oxygen levels fall as plants stop producing oxygen.

Concentration of dissolved oxygen peaks in midafternoon

Respiration also peaks in midafternoon due to all the cellular activity

Dissolved Oxygen ppm

12

10

8

6

4

6:00 am   12:00 pm   6:00 pm   12:00 am

Annual cycle of dissolved oxygen and water temperature in a New England pond

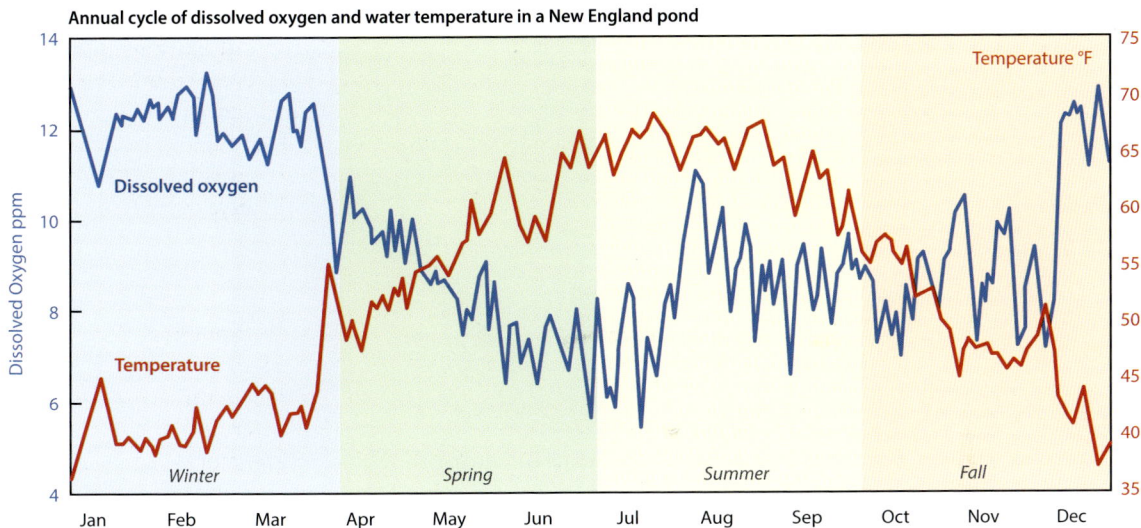

dissolved oxygen levels. In the cooler months, dissolved oxygen levels are higher, and in the warmth of summer and early fall, dissolved oxygen levels are typically at their lowest points of the year.

The close link between water temperature and dissolved oxygen can turn deadly in summer. Small ponds and sunlit rivers can warm to the extent that dissolved oxygen falls to dangerous levels, particularly for fish species. In most areas, the smaller upland streams and rivers typically run through shady forests that help keep waters cool and oxygenated. However, larger rivers and small ponds are usually exposed to summer sunlight that can warm at least the surface water layers beyond what many fish and other aquatic animals can tolerate.

Seasonal cycles

All freshwater bodies in cool temperate environments undergo annual cycles of spring warming, summer heat, fall cooling, and winter

Relationship between dissolved oxygen and water temperature

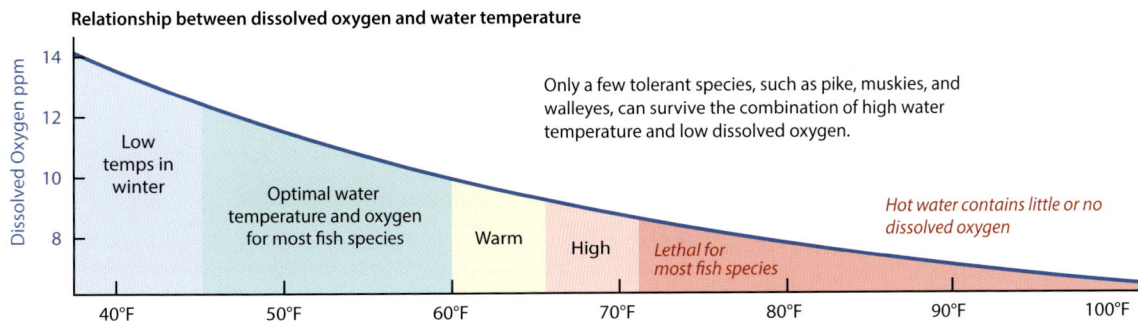

Only a few tolerant species, such as pike, muskies, and walleyes, can survive the combination of high water temperature and low dissolved oxygen.

Low temps in winter

Optimal water temperature and oxygen for most fish species

Warm

High

Lethal for most fish species

Hot water contains little or no dissolved oxygen

SEASONAL THERMAL STRATIFICATION AND CIRCULATION IN PONDS AND LAKES

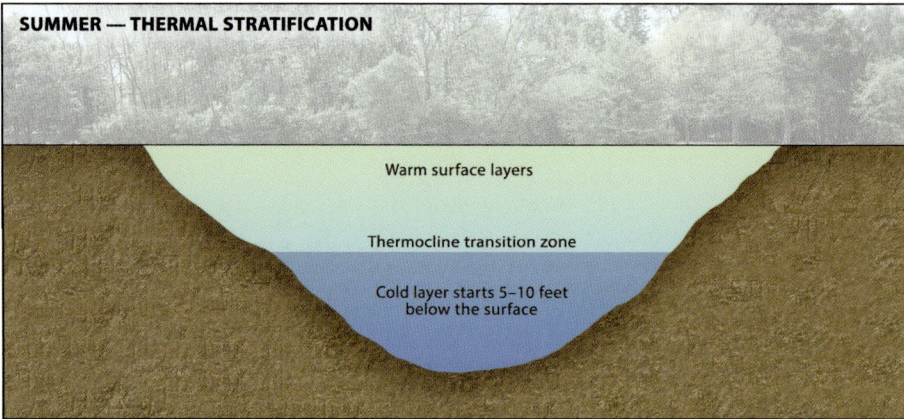

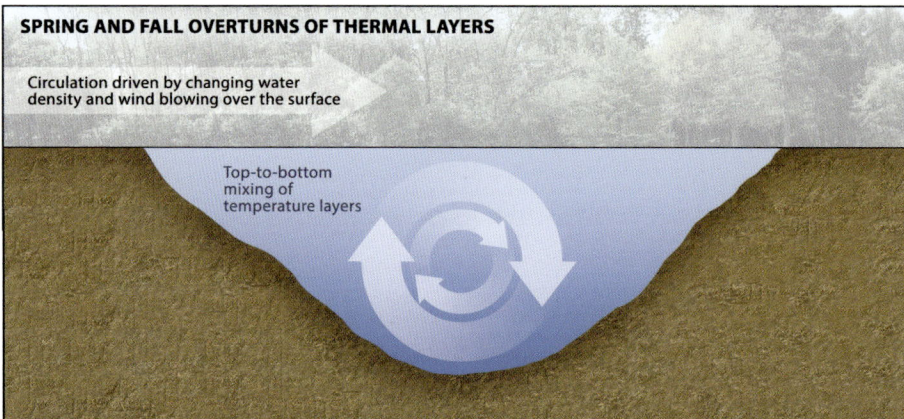

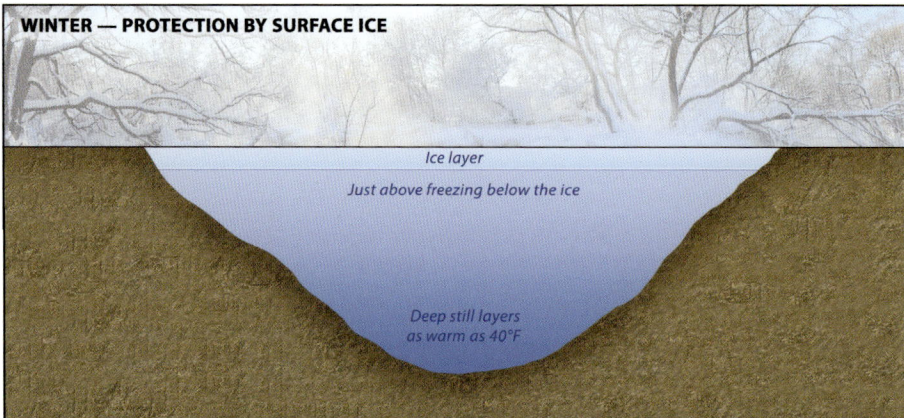

SUMMER — THERMAL STRATIFICATION

Warm surface layers

Thermocline transition zone

Cold layer starts 5–10 feet
below the surface

SPRING AND FALL OVERTURNS OF THERMAL LAYERS

Circulation driven by changing water
density and wind blowing over the surface

Top-to-bottom
mixing of
temperature layers

WINTER — PROTECTION BY SURFACE ICE

Ice layer

Just above freezing below the ice

*Deep still layers
as warm as 40°F*

Seasonal overturn in lakes and ponds is the mixing and circulating of water that occurs seasonally. This cyclical pattern of overturn is caused by temperature and density changes in the water column over a year. Seasonal overturn is critical for ponds and lakes. It helps distribute nutrients and oxygen throughout the water column, providing crucial energy for the ecosystem.

freezing. Although these seasonal changes are primarily temperature cycles, the temperature is intimately linked with water's dissolved oxygen saturation level.

Freshwater ponds are not affected by the Moon or marine tidal movements. Still, ponds—especially larger, deeper ponds—are affected by seasonal temperature and wind variations that drive significant changes in the water column, moving nutrient-rich, deep, cold waters to the surface and drawing surface waters downward to warm and oxygenate the dark pond bottom. Two important physical characteristics of water make pond life possible. The complex crystalline structure of frozen water makes it less dense than cold water, and ice floats. This counterintuitive fact is crucial—otherwise, ice would coat the bottoms of ponds, making it much harder for pond life to survive the winter.

Like most liquids, water becomes denser as it gets colder. But water reaches its highest density at 41 degrees Fahrenheit, still well above freezing. As water temperatures drop below 41 degrees, the near-freezing water is less dense than the waters above it, and the near-freezing water rises to the pond surface to form ice when the temperature reaches 32 degrees. These subtle but essential differences in water temperature and density drive seasonality in pond circulation.

In temperate climates, ponds and lakes typically have two overturning circulation events: spring and autumn. In spring, as the frozen pond surface warms from 32 to 41 degrees, the denser, 41-degree water sinks, displacing the water deeper in the pond. This seasonal circulation is critical to distribute oxygen and nutrients throughout ponds and lakes, which might otherwise become stagnant and lifeless in their bottom waters. In summer, deeper ponds and lakes stratify, with warm surface waters forming a distinctive layer above cold, deep water. In the autumn, the temperature layering begins to collapse as the surface waters cool and sink, and the fall circulation overturn begins.

The mixing and circulating of water in lakes and ponds that occur through the seasons is called seasonal overturn. This cyclical pattern of overturn is caused by temperature and density changes in the water column over a year. Seasonal overturn is critical for ponds and lakes. It helps distribute nutrients and oxygen throughout the water column, providing a crucial energy source for the ecosystem.

Pollution and eutrophication in freshwater systems

Nitrogen and phosphorus are essential but relatively scarce nutrients in pristine natural freshwater systems. Home gardeners will recognize nitrogen and phosphorus as essential elements in plant fertilizers. All living cells need both elements to grow and thrive, and the relative scarcity of nitrogen and phosphorus in clean, unpolluted waters limits aquatic algae and plant growth.

Ice floats because ice is less dense than liquid water. This unusual characteristic of water chemistry makes freshwater environments as we know them possible. The floating ice shields the pond environment and its inhabitants during winter freezes and then disappears when the temperatures warm.

A radiantly healthy pond in late summer. Good pond water is not free from algae; some algae are essential to a balanced and productive pond ecosystem. Here, the water is clear enough to support submerged aquatic plants throughout the pond. This pond is at a well-maintained nature center, so the water is not burdened with excess nitrogen and phosphorus from lawn runoff. In summer and early fall, shallow-water environments such as ponds are always vulnerable to heat and the low-oxygen conditions that warm water brings. Floating water lily leaves provide shadows for fish and other wildlife, but most fish will retreat to the deepest water in the center of the pond to wait out the heat of the day.

Even fortunately located ponds are vulnerable to pollution, however. Streams feed most ponds, and whatever has polluted the stream as it flows over the landscape will end up in the pond. Although few coal-fired plants remain in New England, smoke from distant coal-burning facilities brings a small but constant rain of heavy metals such as mercury and lead, in addition to carbon soot. The phase-out of coal-fired plants has sharply reduced the acid rain and freshwater acidification problems of the 1960s to 1980s, but general air pollution remains a threat to all freshwater environments.

Pond at the Flanders Nature Center, Woodbury, Connecticut.

Eutrophication is a process by which water bodies, including small ponds, become enriched with excess nutrients, usually nitrogen and phosphorus, leading to the overgrowth of aquatic plants and algae. Small ponds are particularly vulnerable. This overabundance of nutrients can cause algal blooms, decreased water clarity, and, most disastrous of all, little or no dissolved oxygen to support healthy aquatic life.

Unfortunately, in today's intensively developed landscape, large amounts of nitrogen and phosphorus commonly enter freshwater in the runoff and stormwater from fertilized lawns, businesses, and farms. Sewage, animal waste, and many home soaps and detergents are other common sources of excess nutrients. These high levels of nutrients drain from developed landscapes during rainstorms and enter ponds, lakes, and rivers, stimulating excess aquatic algae growth.

Excessive algae growth can quickly damage aquatic systems in several ways. Clouds of green algae limit the sunlight submerged aquatic plants need to thrive. Heavy surface mats of algae are unsightly and often smell terrible. They completely disrupt the lives of the many aquatic invertebrates, insects, fish, and other animals that depend on a clean water surface. Worse still, excess algae growth can remove so much dissolved oxygen from the water that fish and other aquatic animals die from a lack of oxygen.

Excess algae growth limits the level of dissolved oxygen in two significant ways. Like all living things, algae cells use oxygen and release

Here a pond within a golf course has been saturated with excess nitrogen, phosphorus, and other artificial nutrients entering the pond from the surrounding lawns. The excess nutrients have caused the explosive growth of cyanobacteria (formerly called blue-green algae). Cyanobacteria form the acid-green mats over this pond's surface and will eventually cover the entire pond. Once a pond is overwhelmed by surface mats, normal subaquatic plant life becomes impossible, and the pond ecosystem begins to die.

carbon dioxide as part of normal cell metabolism (*see illustration, p. 14*). Although green algae produce some oxygen during photosynthesis during the day, excess algae populations revert to oxygen consumption at night. They can quickly use up the dissolved oxygen in the water. As oxygen levels drop, large numbers of algae cells die, and the dead algae tissues then take up yet more dissolved oxygen as the algae decay. This downward spiral of excessive nutrients, the explosive growth of algae, and sharp drops in dissolved oxygen levels is eutrophication, and it can happen in any freshwater or saltwater environment. Unfortunately, most freshwater systems in the Eastern United States show at least some degree of eutrophication from excess nutrient pollution.

Ecological succession and the lifespan of ponds

Ecological succession is a natural process that is so familiar as to be practically invisible. Yet succession is a powerful phenomenon that transforms all environments, both aquatic and terrestrial, over time. In temperate climates along the East Coast, succession transforms barren ground into a young forest in a couple of decades or less (*see illustration, overleaf*).

Succession is the process where the species in a given area change over time, usually in a predictable sequence of plants and animals that

Even ponds in well-maintained parks and nature centers are vulnerable to eutrophication. Parks can work to protect the immediate areas around their ponds from excess runoff of fertilizers, but they can do little about the larger problem of excess nutrients entering the streams that feed the pond from the surrounding suburban and farming landscapes. Once the excess nitrogen and phosphorus enter the pond ecosystem, it is a constant battle to keep algae from overwhelming the natural pond environment.

Given enough time, over many decades most ponds will eventually end up as wet meadows, tree-dominated swamps, or damp hardwood forest areas. This is largely due to the combined powers of ecological succession and erosion. Each stage of pond succession supports a unique community of organisms. The transition from pond to marsh to swamp increases biodiversity by providing diverse habitats for plants and animals. Wetlands such as marshes and swamps also play crucial roles in filtering water, controlling floods, and serving as nurseries for many species.

are dominant for a time and then are outcompeted by other species. Eventually, the result is a stable mix of species, or climax community, that continues in constant form over centuries. In most of the Eastern United States, the climax community around floodplains and more moist uplands is an oak-hickory-maple forest with birches and Black Tupelo scatterings. Live Oak, Red Maple, Black Tupelo, Bald Cypress, Loblolly Pine, Southern Red Oak, and Southern Magnolia dominate the climax moist forest community in the Southeast.

Climax forest communities are the stable endpoint of succession in most of the Eastern United States. Given a few centuries, any barren dirt field in this region will transform itself through succession into a mature hardwood forest.

Succession in ponds and wetlands

Ecological succession also operates in aquatic environments, although the process of aquatic succession is a bit less predictable and visible than the bare-field-to-forest type of succession on land.

Many aquatic successional changes are driven by purely physical phenomena, such as the buildup of silt and plant materials in a pond. As a pond fills with sediment and becomes increasingly shallow, species that favor shallow water will outcompete species that require

Ecological succession over time

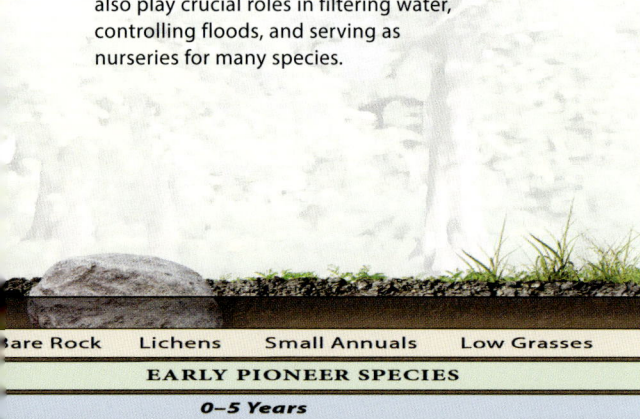

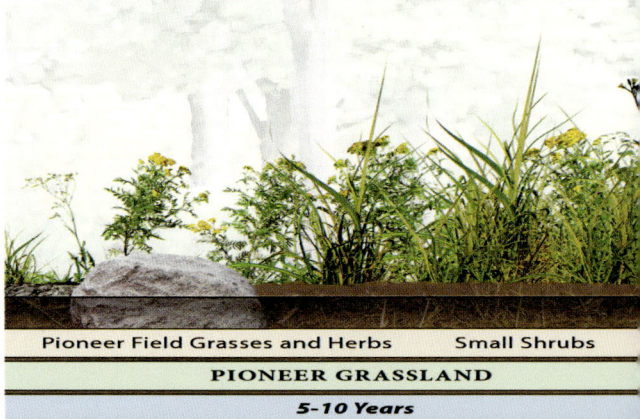

Bare Rock	Lichens	Small Annuals	Low Grasses		Pioneer Field Grasses and Herbs	Small Shrubs
EARLY PIONEER SPECIES					**PIONEER GRASSLAND**	
0–5 Years					*5-10 Years*	

deeper water (*see overleaf*). We tend to notice ecological succession mainly through the changing mix of plant species, but succession also profoundly influences the mix of animal species around freshwater environments. For example, the ubiquitous presence of the White-Tailed Deer in contemporary developed environments is no accident. The White-Tailed Deer is adapted to live along the edges of forests, where open grassy or shrubby communities transition to woodlands. Ecologists call these successional transition zones ecotones. The modern pattern of suburban land development has effectively created a giant forest-edge ecotone that stretches across southern New England, along the Atlantic Coastal Plain, and across the Appalachian Mountain range well into the Midwest.

In the more northern parts of this region, ecological succession in ponds, smaller lakes, and other wetlands is primarily driven by changes in the landscape since the ice of the most recent glacial period disappeared from most of New England 16,000 years ago. As the vast ice sheet melted, countless small ponds and lakes were scattered across a landscape that was largely barren rock and glacial till. Most of these smaller wetlands quickly filled with sediment and, over the past 10,000 years, eventually became forests. Deeper lakes became shallow ponds as they filled with silt, ponds became grass and shrub

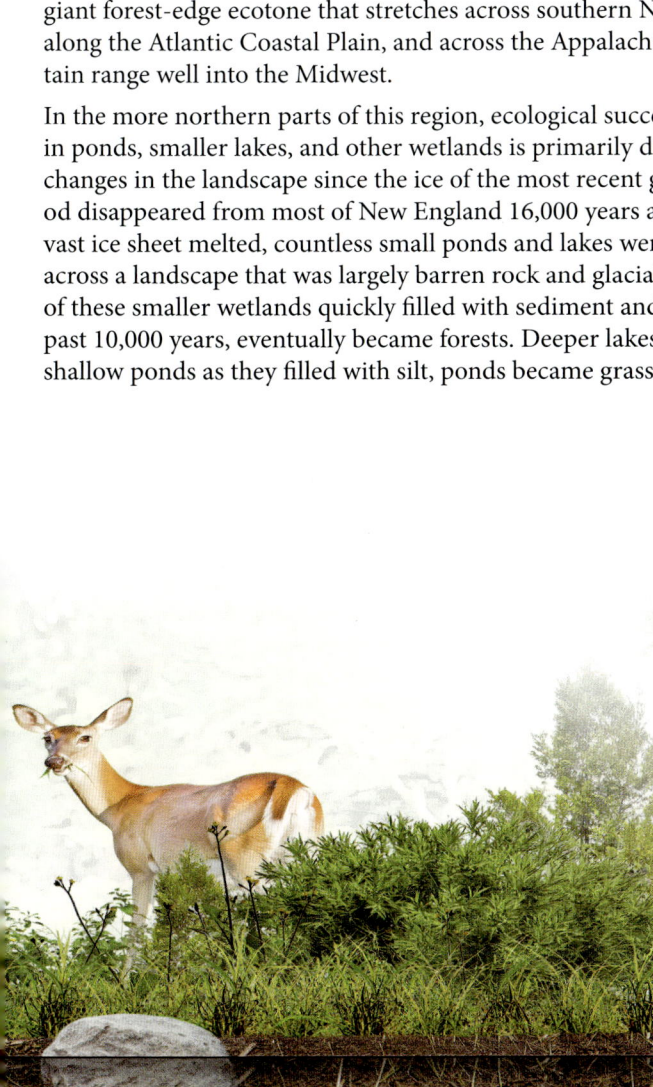

Larger Shrubs and Shade–Intolerant Small Trees

LOW SHRUBS AND MATURE GRASSLAND

10–15 Years

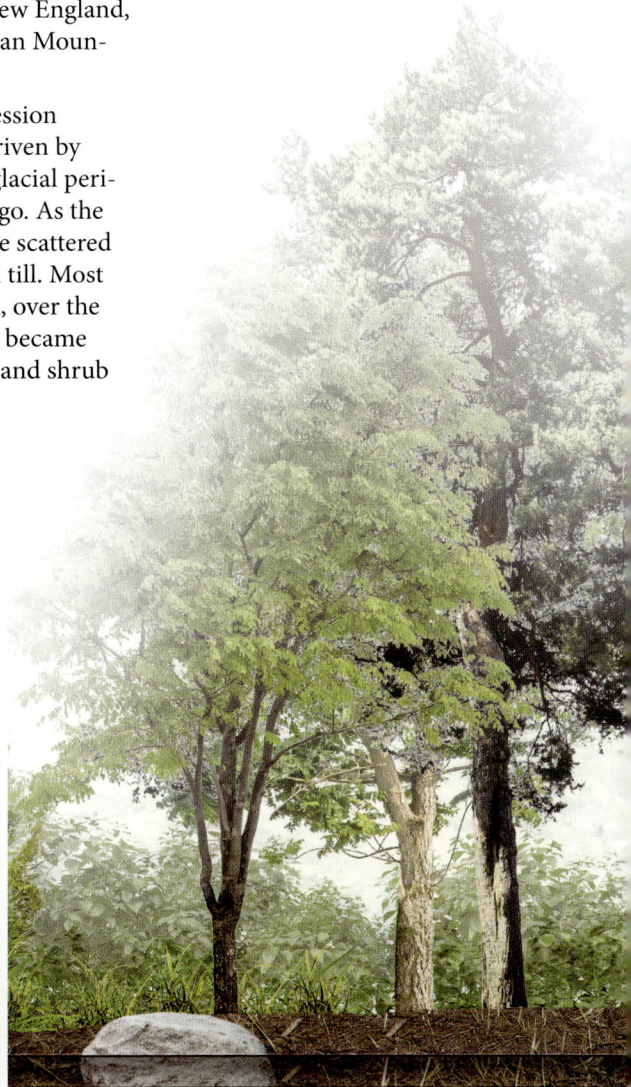

Pioneer Tree Species and Young Forest Trees

PIONEER FOREST SPECIES

15–30 Years

SUCCESSION AND EVOLUTION IN A POND

Ecological succession in freshwater ponds involves a decades-long process in which the pond ecosystem evolves first into a shallow marsh and eventually into a swamp dominated by water-tolerant tree species such as the Red Maple.

1. Pond Stage — Initially, a freshwater pond is typically colonized by aquatic vegetation such as algae and submerged plants. As these plants die and decompose, sediments accumulate at the bottom of the pond, gradually filling it in. This leads to an increase in nutrient levels, promoting further growth of aquatic vegetation.

2. Transition to Marshy Pond — Over time, as the pond fills with sediment, shallower areas begin to support emergent plants such as cattails, sedges, and bulrushes. The accumulation of plant material and further sedimentation continues to fill in the pond, reducing the pond's depth. *See illustrations, pp. 26–27.*

3. Marsh Transitions to Shrub Swamp — The area becomes a marsh, characterized by waterlogged soil and dominant vegetation of grasses and water-tolerant shrub species like Black Willow, Buttonbush, and various alders. Woody plants like shrubs and eventually smaller trees like Red Maples and Black Tupelo trees begin to grow within the layer of shrubs, leading to the development of a swamp. *See illustration, pp. 142–43.*

4. Climax Community — The climax community in this succession process might be a mature tree-dominated swamp or a drier forested wetland, depending on local conditions such as climate and soil type. This stage is relatively stable and can persist until disrupted by environmental change, including climate shifts or human intervention.

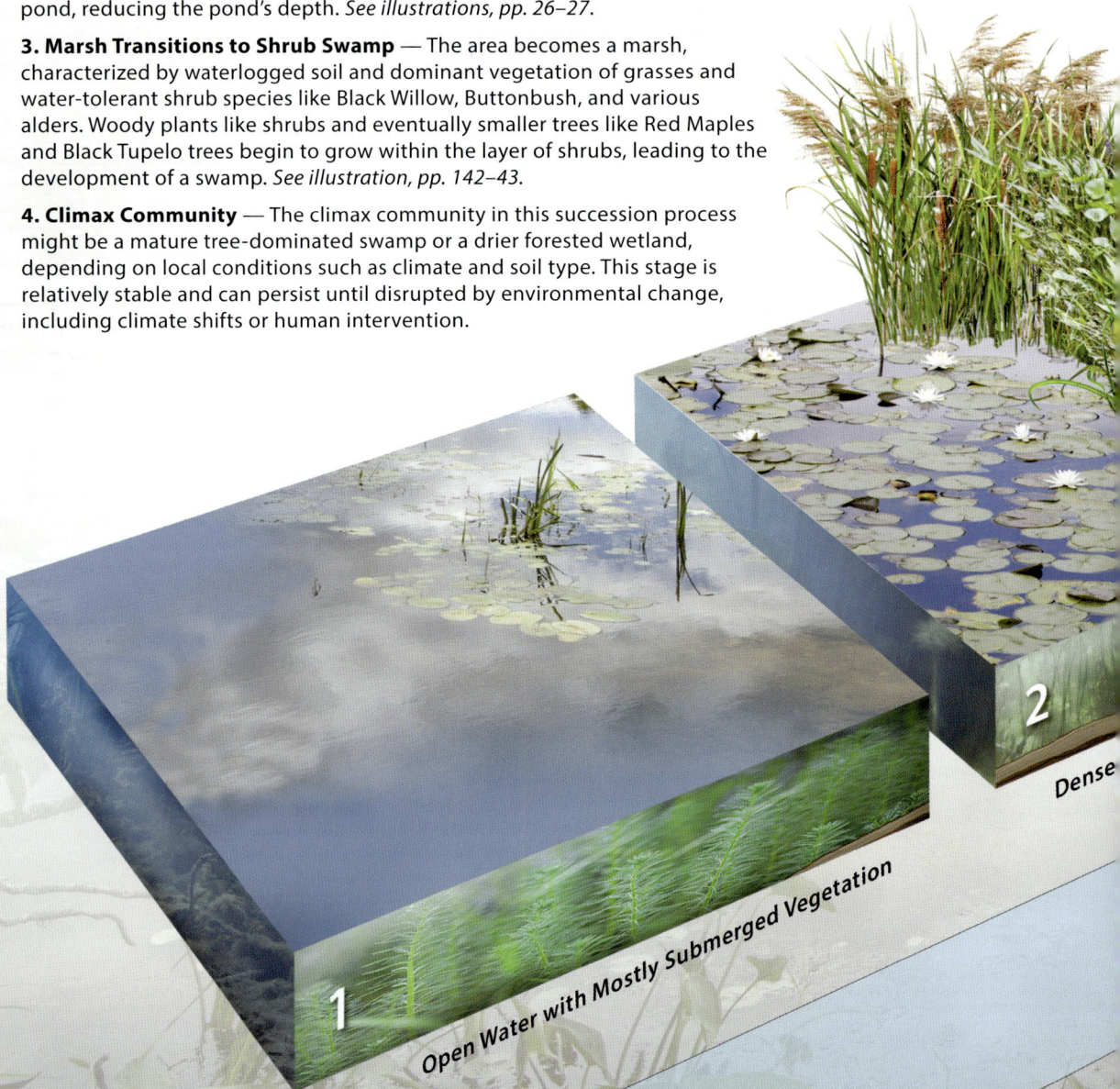

1

Open Water with Mostly Submerged Vegetation

2

Dense

4 *True Swamp, with Hardwood Trees*

3 *Grassy Marsh, Becoming a Shrub Swamp*

Vegetation, Reeds, Sedges

Many decades of time

As sediments fill the entire pond basin, many ponds evolve into shallow marshes or wet meadows.

Here, an old farm pond in Connecticut is filling with silt and now supports wet meadows and grassy marsh plant communities at the right of the image. Given a few more decades, this area will probably dry even more, becoming a tree or shrub swamp and, eventually, the climax forest for the area.

marshes, and the marshes gradually became swamps dominated by water-tolerant hardwood trees such as the Red Maple and Black Tupelo. Over hundreds of years, old drying swamps became forested land. Our natural ponds and lakes are still undergoing a similar, if slower, siltation-driven transition from wetland to dry land.

Human activities have accelerated and distorted long-term succession in our wetlands, often by creating much more erosion that fills wetlands with silt. Land development and farming expose and mobilize large amounts of sediment, which eventually finds its way into local streams and ponds, hastening the filling of all wetlands. But most of the ponds and lakes in our current landscape are not natural. Most of the smaller ponds and lakes we see today are artificial creations of the eighteenth and nineteenth centuries, where small dams were used to create millponds to power grain mills and small factories. Well into the twentieth century, many dam-based ponds we see today were created or maintained solely for aesthetic enjoyment long after the original mills rotted away (*see illustration, p. 8*). Most of the major lakes in the East Coast region are dammed human-made reservoirs designed to give our cities a predictable and safe water supply and sometimes to help with flood control. Because of all the dammed streams, we now have many more freshwater ponds and lakes than we might have otherwise—but there's a tragic irony here. Our many dam-created ponds and lakes were responsible for the death of river migratory fish populations such as the Atlantic Salmon and the American Shad because the dams blocked the breeding streams.

Climate change and ponds

Climate change significantly impacts freshwater ponds in the Eastern United States, altering their ecological balance and threatening the biodiversity they support. Increased air temperatures lead to warmer water temperatures in ponds. This can disrupt the life cycles of aquatic species adapted to cooler conditions. Warmer waters also tend to have lower dissolved oxygen levels, further stressing aquatic life. The increasing heat is a factor in the lifespan of temporary environments such as vernal pools. Pools that once lasted for months may dry in weeks due to increased evaporation.

Climate change is also causing shifts in rainfall patterns, leading to more intense rain events and periods of prolonged drought (*see illustrations, pp. 12, 13*). Heavy rainstorms can increase runoff into ponds, carrying pollutants and excess nutrients from surrounding areas. This can trigger harmful algal blooms, which deplete oxygen and produce toxins, harming fish and other organisms. Conversely, droughts can lower water levels, reducing habitat availability and concentrating pollutants.

The changing and generally warming environment can also play havoc with the timing of annual events such as spring thaws, with consequences for species' mating cycles and normal preparations for winter. Later ice formation and earlier ice-out can disrupt the reproduction and growth of species that rely on specific temperature cues. For example, some amphibians, such as the Wood Frog and American Toad, may breed earlier, exposing their eggs to late-season frosts. Higher temperatures also lead to increased evaporation from ponds. This can lower water levels and increase salinity, particularly in shallow ponds. Increased salinity can negatively impact freshwater species and create conditions favoring salt-tolerant invasive species.

The cumulative effects of these changes pose significant threats to the biodiversity of freshwater ponds in the Eastern United States. Many amphibian species are already showing population declines or shifts in their ranges due to the altered pond conditions. For example, Wood Frogs, Eastern Newts, and Spotted Salamanders (*see illustrations, pp. 114–15, 116–17, 186–87*) have shown population declines in recent decades. Biologists speculate that warming spring weather and a decrease in pond-forming spring runoff from snowmelt may be factors in this decline.

In the later stages of the pond life cycle, any open water disappears and the old pond becomes an emergent marsh and wet meadow, as seen here.

THE FOOD WEB IN PONDS

Even small ponds and lakes are remarkable biological productivity and diversity engines, especially considering that many ponds are frozen or dormant for the winter months. Ponds explode into a complex web of plant and animal interactions each spring using only light and warmth from the sun and water from streams or groundwater.

Primary production in the pond food web is driven by photosynthesis and comprises aquatic green algae, cyanobacteria (formerly called blue-green algae), and larger aquatic plants. Like all green plants, algae and cyanobacteria can produce complex carbohydrates and sugars through photosynthesis. During daylight hours, green plants release oxygen as a byproduct of photosynthesis, increasing the dissolved oxygen levels in ponds. More complex green plants such as grasses, sedges, submerged aquatic plants, and emergent vegetation around pond edges also contribute to the base of the food web (*see illustration, pp. 30–31*).

Many pond and stream animals feed directly on biofilms and aquatic algae, and these primarily tiny aquatic animals and insect larvae form the chief consumer elements of food webs. Larger, more easily seen insects, small fish, and tadpoles are secondary consumers. Finally, larger fish such as sunfish and bass, aquatic birds including the Great Blue Heron and the Osprey, and mammals such as the Raccoon and North American River Otter form the apex or top level of the pond food web.

The microbial loop

In recent decades, biologists and freshwater ecologists have been concerned that the traditional food web, as shown on pages 30–31, did not fully describe the cycling of nutrients in freshwater systems. Well below the level of primary producers (algae and other microscopic green plants), biologists call a deeper level of nutrient cycling the microbial loop.

In every aquatic food chain, there are leftover bits and pieces of organic material that might be scavenged—not living organisms but the organic detritus left behind as organisms are eaten, die, and disintegrate. Looking at pond water under a microscope, you don't just see whole living microorganisms. In between the living animals and algae, you'll see lots of fragments of protein molecules, drops of lipids (fats), bits of carbohydrates, and the cellular innards of previously eaten pond animals floating freely in the water. All this organic debris constitutes a rich soup of nutrients for bacteria, viruses, fungi, and other microorganisms that specialize in feeding on what biologists call particulate organic matter (POM). Think of the organisms of the microbial loop as nature's ultimate cleanup crew for dissolved organic matter.

As the living elements of the microbial loop feed, they, in turn, are eaten by the larger microscopic animals of the pond zooplankton—ciliate protozoans, amoebas, rotifers, and tiny crustaceans such as copepods and *Daphnia*. These animals, in turn, are eaten by small insects such as freshwater shrimp and filtered from the water by mollusks like freshwater clams. Thus, the microbial loop transforms the nutrient

Particulate organic matter (POM) in pond ecosystems is the mix of tiny organic particles suspended in the water. This organic matter consists primarily of decaying plant material, bits of animals, fecal matter, and other biological debris. In ponds, POM is crucial because it serves as a primary source for the recycling of nutrients and energy for various aquatic organisms.

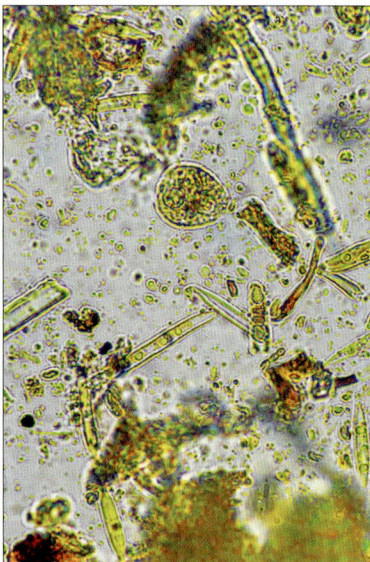

SOLAR ENERGY POWERS THE SYSTEM
Sunlight enables photosynthesis, the primary production of nutrients for the whole pond system.

PHYTOPLANKTON
The Microbial Loop begins with primary production, where photosynthetic organisms such as algae and cyanobacteria use sunlight to convert carbon dioxide and nutrients into organic matter through photosynthesis. These primary producers are the foundation of the aquatic food web, providing energy for higher trophic levels.

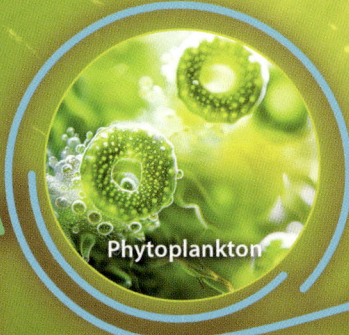

Nutrients enter the higher levels of the food chain

ZOOPLANKTON
The larger animals of the zooplankton are the crucial link to higher levels of the freshwater food chain. Larger zooplankton, such as rotifers and crustaceans, prey upon both bacteria and protozoa. This further transfers the energy and nutrients up the food web.

Phytoplankton

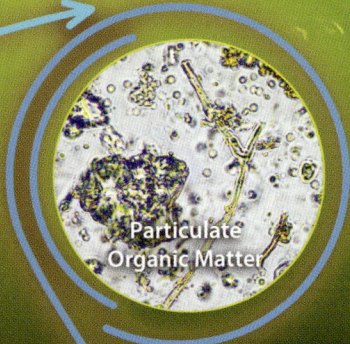

Zooplankton

Particulate Organic Matter

THE MICROBIAL LOOP
Nutrient recovery, regeneration, and energy flow

PARTICULATE ORGANIC MATTER
This includes dead algae, detritus, and other organic debris. POM serves as a substrate for microbial feeding and recovery of the nutrients.

CILIATE PROTOZOANS
Protozoa, including flagellates, ciliates, and amoebae, are microorganisms that feed on bacteria and other small particles suspended in the water. They play a key role in controlling bacterial populations through grazing, thus regulating bacterial decomposition rates and nutrient cycling.

Protozoans

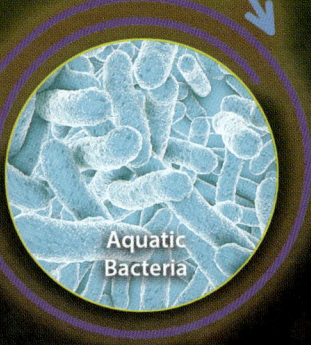

Aquatic Bacteria

BACTERIAL DECOMPOSITION
Bacteria play a crucial role in the microbial loop by decomposing organic matter, including POM. They break down complex organic molecules into simpler compounds. This decomposition releases nutrients such as nitrogen, phosphorus, and carbon back into the water, making them available for primary producers and other organisms.

The Freshwater Microbial Loop
The fundamental base of the pond nutrient cycle

THE FRESHWATER FOOD WEB

Food webs in freshwater ponds are complex networks of interdependent relationships between the different species of plants and animals that inhabit these ecosystems. Food webs in freshwater ponds can also be influenced by other factors, such as the amount of sunlight reaching the water, the presence of pollutants, and the introduction of non-native species. For example, the introduction of non-native fish can disrupt the food web by competing with or predating on native species, which can lead to declines in the populations of native species and changes in the overall structure of the food web.

Biomass of Plants and Animals

NUTRIENT CYCLE
Nitrogen Cycle
Phosphorus Cycle

Decomposition

Nutrient recycling in freshwater ponds is the process by which nutrients are taken up by plants and other organisms, used for growth and metabolism, and then returned to the water column in a form that can be reused by other organisms. This cyclical process is an important component of the overall functioning of freshwater pond ecosystems and helps to maintain the health and productivity of these ecosystems.

Primary producers, such as algae and aquatic plants, use sunlight and dissolved nutrients from the water to produce organic matter through photosynthesis. This organic matter is then consumed by herbivores, such as snails and water fleas, which are in turn consumed by carnivores, such as dragonflies and damselflies. As these organisms grow and reproduce, they release nutrients back into the water column in the form of waste products and dead organic matter. These nutrients can then be taken up again by the primary producers, completing the cycle of nutrient recycling. This cycle is important because it helps to maintain the availability of nutrients in the water, which is critical for the growth and survival of the organisms in the pond.

Animals are not shown to scale

Osprey

Great Blue Heron

Apex Consumers

Pumpkinseed

Secondary Consumers

Common Shiner

Blacknose Dace

Primary Consumers

Daphnia (Water Flea)

Primary Producers

Cyanobacteria   *Volvox* Green Algae

Anabaena Green Algae

Microbial Loop

Water flea
Daphnia sp.

Fall leaves aren't just picturesque—
they power most freshwater ecosystems. The tens of thousands of leaves
that enter the typical pond in autumn
contain a huge load of nutrients. This
bounty is processed by the (mostly
tiny) organisms within biofilms and the
microbial loop, which then becomes
available to plants and animals higher
in the food web.

soup of organic debris into the base of the larger food chain.

Biofilms in freshwater environments

Every surface underwater in ponds, lakes, and other wetlands is
coated with a complex living layer or biofilm composed of green algae,
cyanobacteria, other bacteria, and fungi. Biofilms are hard to see with
the naked eye, but anyone who has ever plucked an old leaf from a
pond or stream has felt the slick, slimy biofilm coating on the decaying leaf. In sunlit areas of aquatic environments, the green algae and
cyanobacteria in biofilms contribute directly to primary food production. However, the bacteria and fungi in biofilms are also critical
in the breakdown of dead plant and animal tissues and recycling of
nutrients in aquatic environments.

Biofilms are also important in the seasonal recycling of pond nutrients. Frost kills off most tiny animals and plant leaves in the autumn.
Biofilm bacteria and fungi break up dead leaves and release their
nutrients into the aquatic environment, aided by the chewing activities of small invertebrate primary consumers such as insect larvae,
isopods, amphipods, and water fleas (*Daphnia*).

Biofilms: A Vital Element of the Freshwater Food Chain
Where the microfauna of the pond break down organic matter

A magnified photo of a biofilm on the surface of a leaf that has been in a pond for at least a few days.

An invisible but important link in the pond food chain

Biofilms are the slimy organic coating that you feel when you pick up a leaf that has fallen into a pond. Biofilms are complex communities of microorganisms, including bacteria, algae, fungi, and protozoa, encased in a self-produced matrix of extracellular polymeric substances (EPS, the slimy stuff). In freshwater systems, biofilms play a crucial role in recycling nutrients like carbon, nitrogen, and phosphorus derived from decaying organic matter. Sticky biofilms also help clean the freshwater system by trapping pollutants, silt, and organic compounds in the water.

Biofilms also support an important freshwater community of protozoans, insects, insect larvae, and other tiny organisms that specialize in grazing rocks and organic matter coated with biofilms. These biofilm grazers in turn form the food for larger insects, fish, and other animals in freshwater systems.

Illustration of a greatly magnified biofilm
Freshwater biofilms are composed of a layer of extracellular polymeric substances (EPS) generated by the bacteria, fungi, and protozoan cells that make up the sticky, slimy layer of the biofilm.

Freshwater ponds are vital ecological habitats that support a diverse range of life-forms and play critical roles in local ecosystems. Ponds serve as breeding grounds and habitats for various species of plants, invertebrates, amphibians, and fish. Healthy ponds contribute to biodiversity by providing different ecological niches and food sources, allowing numerous species to coexist and thrive.

Ponds also play a significant role in the hydrological cycle. Ponds and their associated wetlands act as basins for rainfall and runoff, helping in the recharge of groundwater and moderating the flow of water through landscapes, thereby reducing the risk of floods. In addition, ponds act as natural filters, trapping pollutants and improving water quality by breaking down contaminants through biological processes.

Small freshwater ponds in the Eastern United States face a variety of environmental and pollution challenges. Runoff from urban and agricultural areas introduces excess nutrients, leading to algal blooms that deplete oxygen and harm aquatic life. Pollutants such as pesticides, human pharmaceuticals in wastewater, and heavy metals can accumulate in sediments and contaminate the pond food web. Invasive species disrupt native ecosystems, outcompeting native plants and animals.

THE LIFE IN PONDS

All natural communities are structured in zones that reflect the various physical and biological forces that act on plants and animals to limit or encourage their growth. Physical forces alone, however, do not fully determine the nature of plant and animal communities. The mix of species in a community is also determined by competition for living space and resources among various organisms. The dominant or best-adapted species in the community tend to overwhelm weaker competitors or force them into more marginal living conditions: each ecological zone results from a mix of physical forces and biological competition among the inhabitants.

The plant life in ponds is primarily composed of a mix of submerged, floating, and emergent vegetation. Submerged plants, such as pond-weed and elodea, grow entirely beneath the water's surface, providing oxygen and serving as food and habitat for numerous aquatic animals. Floating plants, such as water lilies and duckweed, are often the most visible plant life, forming mats on the water surface. These floating plants are critical for providing shade, helping to regulate the water temperature, and offering shelter to fish and amphibians. Emergent plants such as cattails and bulrushes extend above the surface at the pond's edges, stabilizing the shoreline and providing nesting sites for waterbirds.

The animal life in ponds is diverse and adapted to the zonal structure of the pond environment. Abundant invertebrates include various insects such as dragonflies and water beetles and crustaceans such as crayfish. These smaller creatures that feed on plants, algae, and each other form the base of the easily visible food web, serving as food for larger predators. At the microscopic level, a far more complex food web exists below what we can see with unaided eyes.

Fish species typically found in these environments include sunfish, bass, and perch, central to the pond's food web dynamics. These fish

Painted Turtles, commonly found in the freshwater ponds of the Northeastern United States, are remarkably adaptable and long-lived. They feed on aquatic vegetation, insects, and small aquatic animals. Painted Turtles can live for 40 years or more. To overwinter, the turtles burrow into the mud and litter at the bottom of the pond, slowing their metabolism drastically to survive the cold months. Painted Turtles are strong, fast swimmers, but you usually see them basking on logs or stones in the pond, using the sun to elevate their body temperature after a cool night underwater.

Great Egret
Ardea alba

Keeping its bottom nesting area clear of Coontail and *Spirogyra* algae is a full-time job for this male Redbreast Sunfish near the margins of a medium-sized pond. The female lays around 3,000 eggs in the nest area. The male then guards the nest and the developing larvae until they hatch after one to two weeks.

prey on smaller animals and are targets for terrestrial and avian predators. Amphibians, such as frogs and salamanders, are also pivotal, using the pond for breeding and feeding. Their presence is often indicated by the chorus of calls during the breeding season, and they play dual roles as both predator and prey within the pond ecosystem.

Birds, too, are attracted to pond waters. Species including kingfishers, herons, and occasionally ducks use ponds as feeding grounds. The presence of such birds highlights the role of ponds and wetlands in connecting aquatic and terrestrial ecosystems.

The biology and ecology of ponds are shaped by seasonal changes, which influence everything from water temperature, ice cover, and dissolved oxygen levels to the breeding cycles of many aquatic and terrestrial species. In the winter, some species hibernate or migrate while others, such as certain fish and amphibians, enter a state of inactivity in the mud at pond bottoms. Spring and summer bring about a burst of life as plants grow, animals reproduce, and migratory birds return.

Typical pond structure and life zones

All Northeastern U.S. ponds, lakes, and river reservoirs behind dams are surrounded by a relatively predictable sequence of aquatic and semiaquatic plants forming distinctive zones (*see illustration overleaf*). These plant zones vary in size depending on the vertical profile of the shorelines. Most mature ponds and lakes have a gently sloping transition between dry forest uplands, an increasingly moist shoreline zone, and shallow areas that typically lie underwater. The edges of some lakes and human-made reservoirs are so steep that the vegetation zone is minimal. Still, most ponds have distinct bands of marshy grasses and shrubs at the edge, emergent wetland grasses, submerged plants beneath the surface, and floating plants in the deeper water.

Redbreast Sunfish
Lepomis auritus

North American Beaver
Castor canadensis

In smaller ponds, the water may be so shallow that the whole pond is vegetated, even in the deeper areas (*see illustration, pp. 46–47*). The small pond in Connecticut's Machimoodus State Park is typical of a central New England pond because it was created with a low dam about a century ago, probably as a farm pond. The pond averages only two to four feet deep and receives sunlight most of the day. Hence, there is a wide variety of emergent vegetation around the open water, and many White and Yellow Water-Lilies are in the deeper water.

Larger ponds and lakes

In addition to rich rims of pond edge vegetation, larger ponds usually show an even more complex mix of submerged aquatic vegetation and a correspondingly wider variety of aquatic animal life (*see illustration, pp. 48–49*). Coontail (Hornwort), Common Elodea, Northern Watermilfoil, and other submerged plants form a complex underwater landscape that offers small aquatic animals food and shelter from predators. Moderate-sized mats of green algae, floating plants such as Common Duckweed and Water Shamrock, and the leaves of water lilies form an underwater maze where tadpoles, young frogs, small turtles, small fish, and many insect larvae can thrive. The structural complexity of emergent, submerged, and floating vegetation is critical protection for young aquatic animals.

As hunting pressure has eased, and wetland protection laws have been developed, North American Beaver activities are once again a significant source of new or enlarged freshwater ponds. Beaver ponds and their surrounding wetlands are highly resistant to wildfires. Their ponds act as natural sponges and temporary reservoirs in times of stormwater flooding. Beaver ponds also store water both aboveground and in the saturated soils around ponds, acting as natural water storage in times of drought.

White Pine

American Beech
Birch species

Idealized cross-section of a pond edge

Black
Gum

Red Maple

Mountain
Laurel

Ferns
Sumacs

Smooth Alder
Arrowwood Viburnum

Goldenrods
Ragweed

Highbush
Blueberry

Local water table

Dry upland forest

Moist Shore Edge

Wet Mead

Upland Zone

Riparian Zone

TYPICAL ZONATION AROUND THE EDGES OF A POND

The typical freshwater pond in the Northeastern United States is surrounded by various plant zones, each characterized by distinct types of vegetation adapted to specific moisture and soil conditions.

Upland Zone — This area is not typically flooded but remains relatively moist. It features a variety of grasses, shrubs, and such tree species as Black Willow, Smooth Alder, Red Maple, and White Pine. This zone acts as a buffer, filtering runoff into the pond and providing habitat diversity and shelter for many common mammal and bird species.

Riparian Zone — This pond border zone is periodically flooded and generally consists of moist but not soaked soils. Moisture-tolerant trees such as Red Maple, Black Willow, and Black Gum are common here. Shrubs including Smooth Alder, Buttonbush, and Arrowwood Viburnum are also common. This zone typically supports a mix of herbaceous plants, including sedges, rushes, and various flowering perennials such as Joe-Pye Weed and Swamp Milkweed (*see illustrations, pp. 94–95*).

Emergent Zone — Located at the pond's edge where the water is shallow. Plants in this zone are usually rooted in the muddy bottom with parts above water. Wetland specialist species such as Common and Narrow-Leaved Cattails, Pickerelweed, Arrow Arum, Tussock Sedge, and bulrushes are typical in this zone, which often has very wet soil or shallow standing water (*see illustrations, pp. 90–91*).

Littoral Zone — This zone includes plants that are submerged or whose leaves float on the water's surface. Common species include water lilies, duckweed, and pondweed. These plants are adapted to grow in water-saturated conditions. Common subaquatic plants that are rooted to the pond bottom include Coontail (Hornwort), watermilfoil, and waterweed (*see illustrations, p. 95*).

Not every pond you encounter will show all of these zones. Depending on the water depth and the extent of very wet soils, you may see ponds bordered closely by wet meadows, extensive grassy marsh areas, shrub swamps, or even tree-dominated swamp areas.

White Water-Lily
Nymphaea alba

Yellow Water-Lily
Nuphar lutea

Common Cattail
Typha latifolia

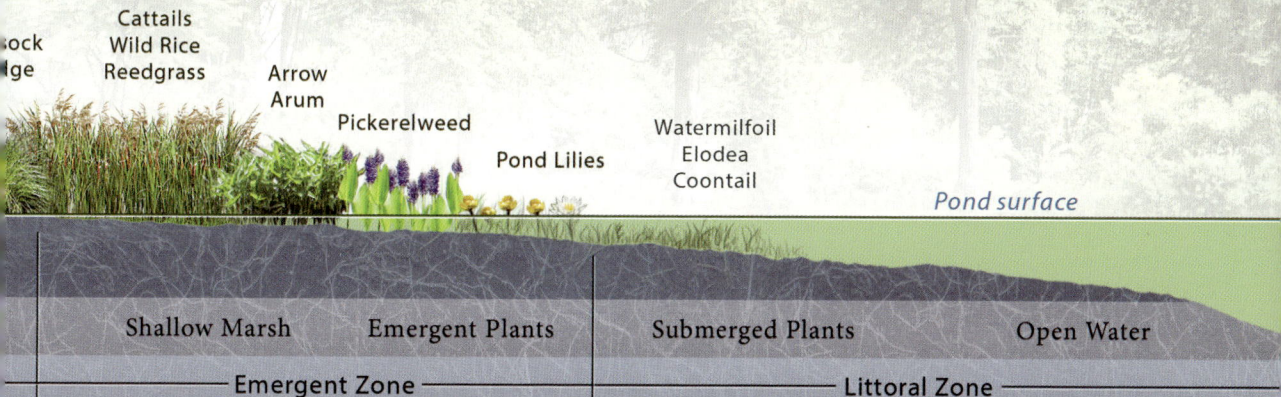

Cattails
Wild Rice
Reedgrass

Arrow
Arum

Pickerelweed

Watermilfoil
Elodea
Coontail

Pond Lilies

ock
ge

Pond surface

Shallow Marsh | Emergent Plants | Submerged Plants | Open Water

— Emergent Zone — | — Littoral Zone —

Small Pond Habitat
Machimoodus State Park, East Haddam, CT

BLUEGILL
4–8 in.

PUMPKINSEED
3–6 in.

Yellow
Water-lilies

**REDBREAST
SUNFISH**
4–6 in.

Giant
Water Bug

Spotted
Turtle

Rusty Crayfish
An invasive species

Slaty Skimmer

Water Strider

Blue Dasher

Belted Kingfisher

Buttonbush

Great Egrets

Arrow Arums

Eastern Painted Turtle

Burr-reeds and other mixed sedge species

American Bullfrog

Northern Watersnake

Bluegills

American Bullfrog tadpoles

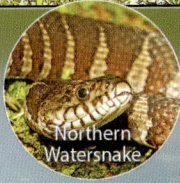

Chain Pickerel
Esox niger

American Mink
Neovison vison

In the open water beyond the sheltered edges of large ponds and small lakes, a whole new complex of fish, turtles, birds, and mammals form the upper reaches of the aquatic food web. Powerful swimmers such as Northern Pike, Largemouth Bass, and Walleyes patrol the transition zones between the pond edge and the deep open water. Chain Pickerels often hide among submerged plants, waiting to ambush unwary small fish, tadpoles, and young ducklings. Larger ponds and lakes offer enough aquatic prey to attract larger predators, including American Mink and North American River Otters. Mink and otters are more common than you might think, but both species are wary and secretive and avoid areas that hikers, fishers, and boaters frequent. In spring and fall migrations, many diving duck species, such as the Bufflehead, use larger ponds and lakes to rest and feed.

The ecology and biology of small ponds in the Northeastern United States can be quite distinct from those of larger ponds and lakes. Small ponds often have less stable environmental conditions compared to larger lakes. Ponds can heat up and cool down more quickly, leading to greater fluctuations in temperature over a day or across seasons. This variability can stress aquatic life that is less tolerant of swings in water temperature and dissolved oxygen, often resulting in a community that either specializes in fluctuating conditions or consists of hardier species. In contrast, large lakes have more buffered thermal and oxygen conditions, supporting a wider variety of aquatic life, including temperature-sensitive species.

Buffleheads are small diving ducks that are common on both freshwater and saltwater ponds and bays. Buffleheads are powerful underwater swimmers that eat insects and plants in freshwater and crustaceans and mollusks in salt water.

The depth of water bodies plays a critical role in oxygen availability, which in turn influences aquatic communities. Small ponds, typically shallower, can undergo rapid oxygen depletion, especially in summer, leading to anoxic conditions that can severely limit the types of organisms that survive there. Larger lakes, with their deeper water columns, generally maintain better oxygen stratification and are less likely to go completely anoxic, supporting more aerobic organisms and complex food webs.

Aquatic plant life also differs significantly between small ponds and large lakes. Small ponds often support a dense growth of aquatic plants and algae throughout, as light penetrates to most of their bottoms. This can lead to eutrophication—where excessive nutrients lead to overgrowth of plants and algae, subsequent oxygen depletion,

Largemouth Bass
Micropterus salmoides

Bufflehead
Bucephala albeola

Pond Aquatic Habitat
Submerged plants and surface vegetation

Watermilfoil
Myriophyllum

Common
Elodea

Hornwort
(Coontail)

Yellow
Water-Lily

Water Clover
(an aquatic fern)

Arrow Arum

White Water-Lily

Bluegills

Watermilfoil
algae

Elodea
algae

Snapping Turtle

Water Boatman

American Bullfrog

Giant Water Bug

Bullfrog tadpoles

Daphnia

CRUSTACEANS

Cyclops

MICROSCOPIC ELEMENTS IN THE POND WATER

ALGAE

Pediastrum

Spirogyra

DIATOMS

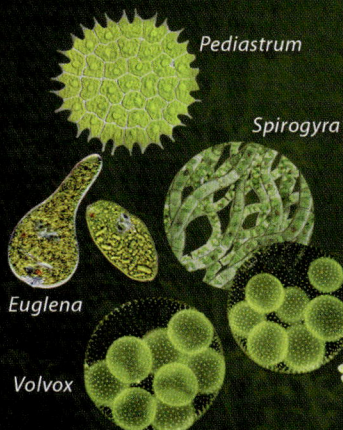

Euglena

Volvox

Closterium

Pond Surface Algal Mat Species

Algal mats on the pond surface

Spirogyra micro view

Closterium algae micro view

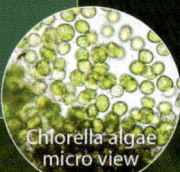
Chlorella algae micro view

Formerly called "blue-green algae"

Cyanobacteria *Gloeotrichia*

Cyanobacteria *Anabaena*

Larger, Deeper Open Water Environments
Large ponds with open water, lakes, and river reservoirs

Bald Eagle

Osprey

Bufflehead

Hooded
Merganser

Common
Goldeneye

River Otters

Largemouth Bass

Northern Pikes

Hooded Merganser

Common Goldeneye

Black Crappie

Yellow Perch

Lake Trout

Chain Pickerel

Walleye

and sometimes fish die-offs. Large lakes, with deeper waters, typically have zoned vegetation, with rooted plants along shallow margins and phytoplankton predominating in open waters, where light does not reach the bottom.

Small ponds usually don't support large or diverse fish populations, especially larger species, due to limited space and oxygen. Smaller ponds may be dominated by amphibians, small fish such as sunfish, and invertebrates such as aquatic insects and mollusks. Larger lakes, thanks to their greater volume, depth, and food availability, can support larger fish species and a more complex food web, including top predators such as large fish and predatory birds.

The sheer surface size of larger ponds and lakes also influences the kinds and numbers of birds you might see. In a smaller pond, with pond edges always nearby, aquatic birds are in constant danger from land predators including raccoons, coyotes, and bobcats. Larger ponds and lakes can safely host large flocks of birds that float on the surface far beyond the easy reach of land predators. Large ponds and lakes are often critical habitat stops for migrating aquatic birds, offering both food and refuge from most predators.

Larger ponds and lakes provide protection for large flocks of waterbirds—the sheer distance from shore discourages most land-based predators. Larger lakes and ponds also offer more food resources during migration periods. Ponds such as New Hampshire's Pondicherry ponds (*see illustration, pp. 198–99*), Massachusetts' Quabbin Reservoir, and the many ponds of New Jersey's Edwin B. Forsythe National Wildlife Refuge (*see illustration, pp. 234–35*) are critical rest and feeding stations for bird migration in the Eastern United States.

Adaptations to life in water

Plants and animals living in freshwater environments such as ponds face several common challenges.

Osmotic balance between water and salts

Freshwater organisms have cells with higher salt concentrations than the surrounding water. This causes water to move into the organism through osmosis and might lead cells to swell and burst unless the osmosis is actively managed by cell physiology. The biochemistry of this active salt-and-water balance maintenance is called osmoregulation. Plants have developed strong cell walls resistant to osmosis and have specialized structures such as contractile vacuoles that pump excess water out of cells. Freshwater animals such as fish actively regulate the salt and water balance in their bodies. Aquatic animals have specialized organs such as gills and kidneys that filter out excess water and maintain proper salt concentrations. Some freshwater animals have developed scales, shells, or mucus coatings that help prevent excessive water absorption and protect them from osmotic stress.

Oxygen for respiration

All plant and animal cells require oxygen to function normally. Plant cells typically absorb dissolved oxygen in water directly through their cell walls. Dissolved oxygen levels in water are much lower than in air, so larger aquatic animals such as fish have gills that are highly efficient at extracting oxygen from water. Animal circulatory systems then distribute oxygen from the gills to the rest of the body. Tiny insects and younger frog tadpoles don't require an active circulatory system and can absorb sufficient dissolved oxygen directly through their body surfaces.

Temperature regulation

In aquatic environments, there is a tight relation between water temperature and dissolved oxygen. Warm water can't hold as much dissolved oxygen as cold water, so seasonal water temperature changes directly affect how plants and animals can obtain oxygen.

Water is dense and has a great capacity to absorb and store heat. This resistance to rapid temperature change is called thermal inertia. Water temperature changes more slowly than air temperature, and this thermal inertia helps to buffer the aquatic environment from sudden weather changes.

This slowness to change temperature can have good and bad effects on aquatic animals. In the heat of summer, pond water absorbs heat and can retain that heat well after the sun sets. Warm water contains little dissolved oxygen, and prolonged oxygen deficiencies can damage or kill animals and plants. Aquatic animals are generally ectothermic (relying on external heat sources), and their activity levels and

American Bullfrog tadpole
Lithobates catesbeianus

Backswimmer
Family Notonectidae

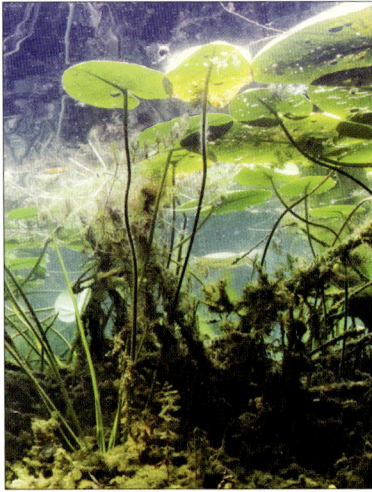

Water lilies and aquatic plants stay buoyant thanks to aerenchyma tissue, a specialized spongy tissue filled with large air spaces. These interconnected chambers run throughout the stems and leaves, providing both buoyancy and a pathway for oxygen to reach underwater parts of the plant through diffusion.

metabolic rates are adjusted to the temperature of their environment. Colder water can slow the metabolism of ectothermic (cold-blooded) animals, but oxygen usually is plentiful in cold water, so many animals such as sunfish prefer the deeper, cooler waters of ponds on hot days.

Buoyancy

Gravity operates more slowly in the density of water, but without some means to create neutral buoyancy, all plants and animals would eventually sink to the bottom of the pond. If you've ever tried to retrieve an object from the deep end of a swimming pool, you'll know that it takes a lot of energy to swim downward when your body is buoyant because your lungs are full of air. This is why human scuba and snorkel divers usually wear weight belts to help achieve neutral buoyancy, so the divers don't waste a lot of swimming energy just trying to stay underwater.

Many aquatic plants have air-filled tissue (aerenchyma) that helps them float near the water surface and maintain access to the sunlight necessary for photosynthesis. The mud and leaf litter at the bottom of ponds contains little oxygen. The porous aerenchyma tissue in aquatic plant stems and roots helps the plant distribute gases to roots that would otherwise be starved of oxygen.

Some small insects capture bubbles of air that supply oxygen while underwater, and the air bubble also helps them maintain neutral buoyancy. Fishing spiders, giant water bugs, and other pond predators capture air bubbles at the surface before diving to seek prey. These captured air bubbles act like miniature scuba tanks for these active diving hunters. Common surface insects such as whirligig beetles capture bubbles of air to help keep them afloat on the pond surface. The fine hairs on the legs and feet of water striders and fishing spiders also capture microbubbles of air that allow the animals to skate over the pond's surface without breaking the surface tension of the water.

Fish have swim bladders that help them maintain neutral buoyancy. A swim bladder is an internal gas-filled organ that helps the fish keep its buoyancy, allowing it to stay at its current water depth without expending energy in swimming. The swim bladder can be filled or emptied through gas exchange with the bloodstream, depending on the fish's needs. For buoyancy control, a fish decreases the density of its body by increasing the volume of gas in its swim bladder, making it more buoyant. To sink, the fish releases gas from the swim bladder, reducing the bladder volume and increasing the body's overall density.

Movement through water

Water is a very dense environment for a small aquatic animal, as thick as molasses to a small fish or insect. Active swimmers are often streamlined for easier movement through water. Fish, for instance, have bodies shaped to minimize drag. Aquatic animals also have fins,

ADAPTATIONS TO FRESHWATER ENVIRONMENTS

The effect of osmosis on cells in water

Saltwater Conditions

HYPERTONIC
Water flows out of cells and the cells shrink

Balanced Conditions

ISOTONIC
Water flows into and out of cells equally

Freshwater Conditions

HYPOTONIC
Water flows into cells and the cells fill and enlarge

Fish Adaptations to Freshwater and Buoyancy

The stomach actively absorbs salts from food

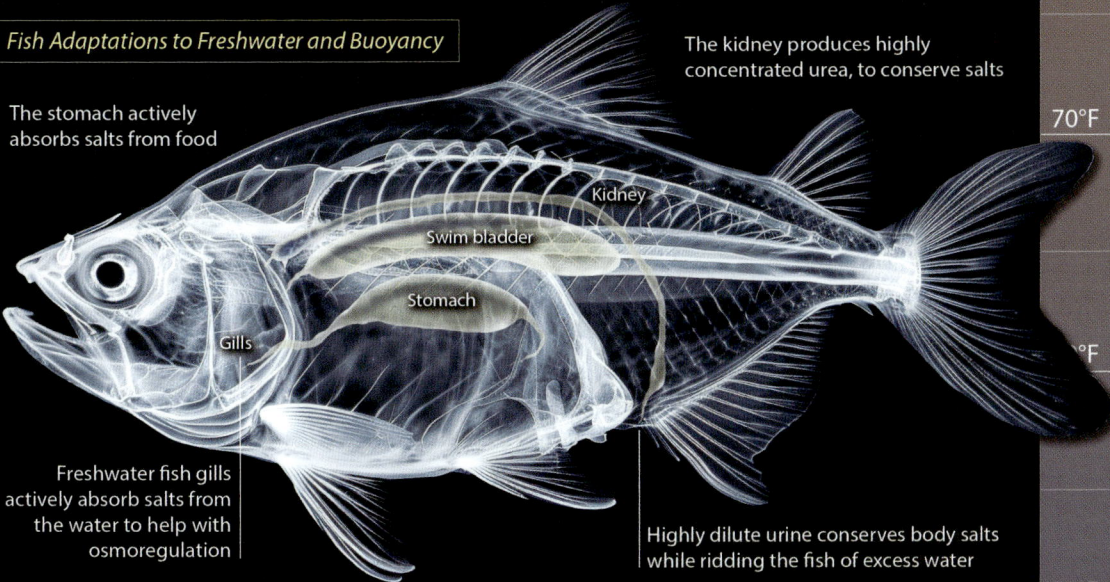

The kidney produces highly concentrated urea, to conserve salts

Kidney

Swim bladder

Stomach

Gills

Freshwater fish gills actively absorb salts from the water to help with osmoregulation

Highly dilute urine conserves body salts while ridding the fish of excess water

Lateral line senses vibration and movement in the water

ADAPTATIONS

Osmoregulation: To maintain balance, fish actively absorb salts through their gills and skin while excreting large amounts of very dilute urine to expel excess water.

Respiratory: Fish have adapted efficient respiratory systems to maximize oxygen uptake through their gills.

Camouflage and defense: Freshwater fish often have coloring and patterns that help them blend into their environments to avoid predators or to avoid warning prey animals of their approach.

Locomotion: The structure of fins and body shapes in freshwater fish can be adapted to different types of water currents. Fish in ponds are often more laterally flattened for easier maneuvering in tight spaces.

The scales and mucus coating on freshwater fish are crucial to resist the absorption of excess water. The mucus also helps the fish ward off infections and parasites and reduces friction while swimming.

O₂

Temperature	O₂
90°F	2 ppm
80°F	4 ppm
70°F	6 ppm
	8 ppm
50°F	10 ppm
40°F	12 ppm
30°F	14 ppm

Lethal heat and lack of oxygen (hypoxia)

Ideal temperatures and oxygen levels

Adaptations to winter conditions

Water boatman
Family Corixidae

Backswimmer
Family Notonectidae

webbed feet, and mobile tails that enable precise movements, efficient paddling, and speed control. Active swimming insects such as giant water bugs, backswimmers, and water boatmen also have streamlined bodies and flattened legs for more efficient swimming. Pond animals such as dragonfly larvae are less streamlined for movement because they are primarily ambush hunters, not active swimming predators.

Sensory adaptations in animals

Water can limit visibility and hearing. Freshwater animals may rely more on other senses, such as vibrations (detected through lateral lines in fish) or chemical cues (smell and taste), to navigate, communicate, and find food. Humans have excellent eyesight, but most aquatic animals and insects that live around ponds have much more limited vision. Fish and active flying insects such as dragonflies and butterflies have surprisingly limited eyesight compared to ours but are very sensitive to sudden movements in their field of view. Remembering that aquatic animals are sensitive to vibrations and sudden movements can help you get much better looks at pond life. Be quiet, use light footfalls, and move slowly and deliberately when approaching a pond or lake edge.

Inspect the pond edges as you approach the shoreline. Many frogs and smaller birds use the shelter of shore vegetation to scan the water for food and prey. Look closely at the shadowed areas under pondside grasses and sedges. Go slow, and avoid the urge to chat with companions. Try this stealthy approach to watching wildlife, and you'll be delighted at how much more you will see.

BELTED KINGFISHER (female)
Megaceryle alcyon

The Belted Kingfisher is a striking bird commonly found across the freshwater ponds and lakes of the Northeastern United States. Kingfishers are recognizable by their blue-gray plumage, shaggy crest, and characteristic white collar around their neck.

Kingfishers are noisy birds whose rattling calls are among the most frequently heard sounds around freshwater wetlands. Kingfishers are one of the few bird species that can genuinely hover in midair. They use this hovering skill above the water surface while hunting small freshwater fish. When the kingfisher spots a fish, the bird collapses its wings, dramatically dives through the water's surface, and grabs the fish with its beak.

Surface tension: A unique property of water

The surface layer of a pond is a unique and dynamic environment where water's physical and chemical properties interact to create a habitat for a diverse array of organisms. Biologists call this area of the pond the neuston, and it is particularly well suited for small, light-weight animals such as water striders (family Gerridae) and whirligig beetles (family Gyrinidae) that have evolved remarkable adaptations to exploit the pond surface (*see illustration, overleaf*).

The physics of surface tension

Surface tension is the physical force that allows these animals to remain on or near the surface. Water molecules (H_2O) are like weak magnets, with a positive end of the oxygen atom and a negative end in the two hydrogen atoms. Although the magnetic forces involved are

The Neuston Zone of Lakes and Ponds
Animals that live at or near the surface of the water

Fishing Spider
Dolomedes

Whirligig Beetles
Gyrinidae

Dragonfly
larva

Giant Water Bug

The undersurface of the water is also
barrier, and a danger to small fish tryi
to escape predators such as pikes,
pickerels, and the Largemouth Bass.
The predators will drive the tiny fish
upward toward the surface, where th
can be trapped and picked off one by
one, because smaller fish aren't stron
enough to jump through the
surface tension.

Many types of insect predators lurk
just under the water surface.

Largemouth
Bass

Mosquito larvae live on the
undersurface of very still
ponds. They hang upside
down and breathe through a
siphon tube that breaks the
water surface. The pond
surface provides the larvae
with access to oxygen and a
source of food, such as the
algae and microorganisms
near the pond surface.

The air just above pond surfaces is dense with many types of small flying insects. Mosquitoes, flies, mayflies, damselflies, and dragonflies all make rich pickings for expert aerial predators such as Barn, Cliff, and Tree Swallows.

Barn Swallow

Water Strider
Gerridae

The molecules at the surface have fewer neighboring molecules to interact with, resulting in an imbalance of forces. The net force on a surface molecule is directed inward, toward the bulk of the liquid. This inward force creates a "skin" on the surface of the liquid, known as surface tension.

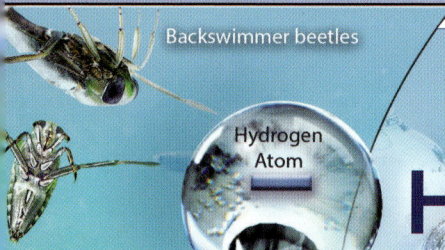

Backswimmer beetles

The surface line of molecules is like a chain of magnets, maintaining the surface tension

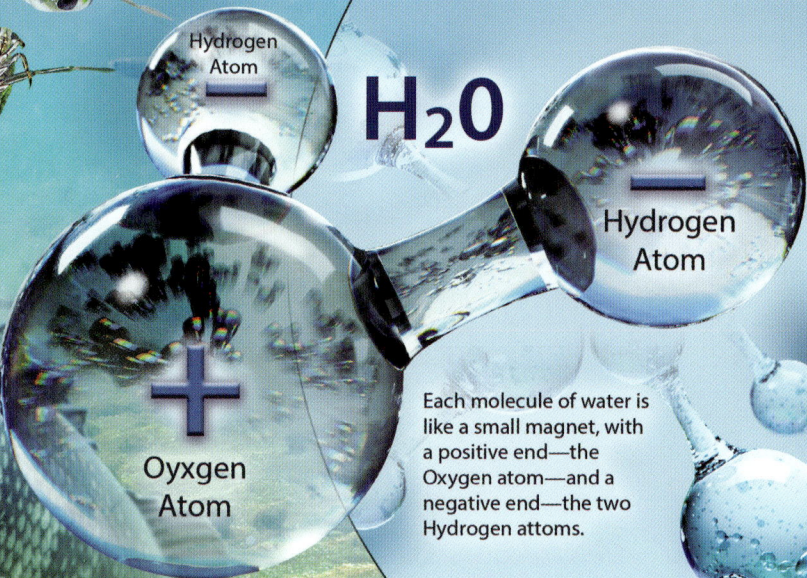

Hydrogen
Atom

H_2O

Hydrogen
Atom

Surface tension in water comes from the cohesive forces between the molecules. Below the surface, each molecule is surrounded by other molecules and experiences attractive forces from all directions. These forces balance out, resulting in no net force on the molecule.

Oyxgen
Atom

Each molecule of water is like a small magnet, with a positive end—the Oxygen atom—and a negative end—the two Hydrogen attoms.

weak, at the interface between water and air, the cohesive forces between water molecules are stronger than those between water and air molecules. This creates a thin, elastic layer at the surface, analogous to a stretched membrane. With strong hydrogen-oxygen bonds, water has a relatively high surface tension. This cohesive force is sufficient to support small objects if their weight is distributed over a large enough area. These are the same forces that cause water to bead on a smooth surface. Water molecules stick to each other far more than to other substances.

The chemical properties of the surface layer play a crucial role in the ecology of the pond neuston region. The surface microlayer is rich in organic matter, including bacteria, algae, and pollen. This layer acts as a food source for many surface-dwelling and subsurface-dwelling organisms. You can easily see this complex layer of surface nutrients on spring days when trees release large amounts of pollen into the air above the pond. Much of the pollen ends up on the pond surface, enriching the neuston and its animals.

Adaptations of water striders

Water striders are the quintessential inhabitants of the water surface. Their long, slender legs distribute their weight evenly, preventing them from breaking through the surface tension. These legs are also covered in microscopic hydrophobic (water-repelling) hairs that trap

The key to the water strider's ability to stay on the pond surface lies in surface tension, a property of water in which molecules at the surface are more strongly attracted to each other than to the air above, creating a slight "skin" on the water's surface. Water striders take advantage of this phenomenon through several adaptations.

First, water strider legs are covered in thousands of tiny hydrophobic (water-repelling) hairs called microsetae. These hairs trap tiny air bubbles, which increase buoyancy and reduce the legs' likelihood of breaking the water's surface tension. Second, the legs are long and slender, distributing the insect's weight over a larger area.

When a water strider's leg touches the water, it creates a slight depression without breaking the surface. This depression increases the surface area of water in contact with the leg, providing more support. The insect's lightweight body also helps it stay afloat.

air bubbles, further increasing their buoyancy. The movement of water striders is a fascinating example of fluid dynamics. They propel themselves by creating tiny ripples with their middle legs, which act like oars. These ripples generate vortices in the water, which the striders exploit for forward motion. Their movements are so efficient that they can reach a speed of up to three miles per hour.

Beetles in the neuston region

While water striders remain on the surface, many beetles inhabit the subsurface layer beneath the surface tension. Beetles in this layer employ a variety of strategies to maintain their position. Some, like whirligig beetles (family Gyrinidae), have a unique body shape with a divided eye that allows them to see both above and below the water. Like water striders, these beetles have specialized legs for swimming and creating surface ripples. Whirligig beetles also carry a tiny bubble of air at the end of their body, supplying oxygen and buoyancy to the fast-moving animals.

Other beetles, such as backswimmers (family Notonectidae), use hydrophobic hairs on their abdomen to trap a layer of air as a buoyancy aid. They swim upside down, using their long hind legs for propulsion. The upside-down orientation allows backswimmers to see and capture prey at the pond surface easily.

Even relatively large animals can exploit the surface tension of ponds. Fishing spiders (*Dolomedes sp.*) have legs that are dense with hydrophobic hairs, which trap tiny air bubbles that create buoyancy. Fishing spiders are strong enough to break the surface tension of the water and dive after prey below the surface.

Whirligig beetles (family Gyrinidae) are able to stay on the surface of freshwater ponds due to a combination of physical adaptations. Their oval and flattened body shape allows them to float easily. A waxy, water-repellent coating on their body helps them resist sinking. And their specially adapted middle and hind legs, which are flattened and paddle-like, allow them to skate across the water's surface. These legs also have tiny, hydrophobic hairs that trap air bubbles, further enhancing the beetles' buoyancy.

The radical differences in life habits, prey items, and even basic body form and shape of larvae versus adult dragonflies is no accident. Dragonfly larvae are fierce aquatic predators in ponds. Adult dragonflies are masters of the air over ponds and wetlands.

If you look closely at pondside vegetation, you'll often see the empty larval shells left behind by dragonflies that have emerged as adults. In this image, two larvae are clinging to a vertical leaf during the final process of metamorphosis.

Metamorphosis as a life strategy

Many of the most common pond animals have evolved complex life cycles that take advantage of the very different environments of water, the solid pond edges, and the airspace around the pond. Metamorphosis is the general process that allows animals as different as flies, dragonflies, frogs and toads, and salamanders to live at least part of their lives in very different forms, in the highly varied environments of water and air. Metamorphosis offers organisms a distinct survival advantage by dividing resources and habitats across life stages. This separation reduces competition between juveniles and adults for food, space, and other essential resources. Metamorphosis also allows for specialization, with larvae often adapted to exploit different food sources or microhabitats than adults, further enhancing survival and growth potential. For instance, American Bullfrog tadpoles are primarily herbivorous, feeding on algae and aquatic plants, while adult American Bullfrogs are carnivorous, preying on insects and other invertebrates.

The dragonfly populations of small ponds make an ideal and easily observable cycle of metamorphosis in organisms. The larger dragonfly species life cycle begins with egg laying (*see illustrations, opposite and overleaf*). Female dragonflies lay their eggs in or near water bodies, including small ponds, lakes, and marshes. They often insert their eggs into plant tissues or drop them onto the water's surface. The eggs are small and are typically laid in batches. The egg-laying stage is crucial in determining the future population density and distribution of the species within a habitat.

After the eggs hatch, the life cycle progresses to the larval stage, which occurs entirely underwater and can last from several months to a couple of years, depending on environmental conditions such as temperature and food availability. Dragonfly larvae, or nymphs, are voracious underwater predators, equipped with a unique lower lip, or labium, which they can extend forward to snatch prey. They feed on a wide variety of aquatic organisms, including smaller insects, tadpoles, and even small fish. This predatory behavior is essential for controlling the population of other species in smaller ponds, contributing to an ecological balance in an environment with limited resources and living space.

The transformation from nymph to adult occurs in distinct steps. As the nymph grows, it goes through several molts, shedding its exoskeleton multiple times as it increases in size. The final stage of the nymph's life sees it leave the water. This emergence is typically synchronized with favorable weather conditions. The nymph climbs up a reed or other suitable vegetation protruding from the water and begins the transition to adulthood.

During metamorphosis, the dragonfly's body undergoes dramatic changes. The nymphal skin splits along the back, and the adult dragonfly pulls itself out of its old exoskeleton. The newly emerged adult is soft and highly vulnerable to predation. It takes some time for the dragonfly's wings to harden and its body to gain the colors characteristic of its species. This drying and hardening process is crucial for survival, enabling the dragonfly to fly and hunt effectively.

Once matured, the adult dragonfly spends its time feeding and reproducing. Mating dragonflies are easy to spot near ponds in mid- to late summer. The mating male and female pairs lock together, sometimes in flight and sometimes on floating vegetation at the pond surface. Adult dragonflies are powerful fliers, capable of capturing prey midair, and are known to migrate over considerable distances. The adults typically live a few weeks to a few months, during which they must reproduce to ensure the continuation of their species.

The Green Darner (*Anax junius*) lays its eggs on aquatic vegetation or directly into the water. The female inserts her ovipositor into plant tissue or drops eggs at the water's surface, as shown below. She often deposits eggs while the male is still grasping her, a position called tandem oviposition (*see illustration, overleaf, no. 1*). The elongated eggs typically hatch in one to five weeks, depending on water temperature. Once hatched, the aquatic nymphs (also called naiads) emerge and begin their underwater life stage.

The Life Cycle of Dragonflies
Fierce predators both above and below the pond surface

EASTERN PONDHAWK (Female)
Erythemis simplicicollis

The Eastern Pondhawk is an aggressive perch-hunting dragonfly that hunts flying insects over and around freshwater ponds and streams. While deadly to gnats and flies, dragonflies are entirely harmless to people.

FINAL INSTAR STAGE
After a series of molts and a period of growth, the nymph climbs out of the water onto emergent vegetation.

MATING AND EGG-LAYING
The life cycle of a dragonfly begins with oviposition, where the female dragonfly lays eggs. She typically deposits her eggs either in or near water, as most dragonfly species have aquatic larvae.

1

3

Dragonfly Nymph

Dragonfly Nymph

Dragonfly nymphs typically eat other aquatic insects, but the larvae of larger dragonflies can catch tadpoles and small fish as well.

TENERAL STAGE
Upon emerging from the water, the nymph goes through a brief period called teneral stage. The newly emerged adult is soft and vulnerable as its exoskeleton hardens. It remains near the emergence site while its wings expand and dry.

LATE TENERAL AND ADULT STAGE
Once the wings have fully expanded and dried, the dragonfly becomes a mature adult, also known as an imago. The dragonfly takes to the air, to feed on flying insects, and to find a mate.

Animals are not shown to scale

The primary goal of adult dragonflies is to mate and reproduce. Dragonflies use visual and auditory signals to attract potential mates. Once mating occurs, the female dragonfly will lay her eggs in or near water, starting the cycle anew. This reproductive cycle repeats a number of times each summer, as dragonflies typically live for several weeks to a few months, depending on the species, environmental conditions, and predators. In the autumn the final annual generation of the aquatic nymph stage will overwinter deep in the litter at the bottoms of ponds and streams, to emerge as new adults in late spring.

AQUATIC NYMPH STAGE
Nymphs are aquatic and live underwater in freshwater habitats like ponds, lakes, or slow-moving streams. They have elongated bodies with six legs and are voracious predators, feeding on small aquatic organisms such as tadpoles, mosquito larvae, and small fish.

Emerald Spreadwing
Lestes dryas

Spatterdock Darner
Rhionaeschna mutata

Eastern Forktail
Ischnura verticalis

Fragile Forktail
Ischnura posita

Distinct dragonfly communities in wetland habitats

Animals are not randomly distributed across wetland landscapes. Every animal you see is keenly adapted to a specific environment, ecological competitors, and types of prey available. Wetland habitats can show remarkable biodiversity, with dragonflies and damselflies (order Odonata) serving as both indicator species and easily observed charismatic animals. These insects exhibit highly specific habitat preferences, resulting in distinct communities across various freshwater environments. Understanding the distribution and composition of Odonata populations can give you valuable insights into the health and characteristics of each kind of wetland habitat you visit.

Vernal pools

Vernal pools—ephemeral water bodies that typically fill with spring rains and dry out by summer—host unique dragonfly communities adapted to their temporary nature. These habitats are characterized by the absence of fish predators, allowing for the proliferation of specialized Odonata species. In vernal pools, salamanders frequently occupy the role of top predator, influencing the composition of the dragonfly community. Species such as the Emerald Spreadwing (*Lestes dryas*) and the Spatterdock Darner (*Rhionaeschna mutata*) are often associated with vernal pools. These species have rapid larval development cycles, enabling them to complete their life cycles before the pools dry out in summer.

Seasonal ponds without fish

Small, seasonal ponds present a different ecological niche, often dominated by damselflies as the top invertebrate predators. These habitats support species such as the Eastern Forktail (*Ischnura verticalis*) and the Fragile Forktail (*Ischnura posita*). The presence of these smaller Odonata as top predators can be attributed to the limited size and depth of these water bodies, which usually don't support larger dragonfly species or fish populations.

Larger fish-dominated ponds

Larger ponds and small lakes present more complex ecosystems where larger dragonfly species and pond fish assume the role of top predators. These habitats support a diverse array of larger dragonflies, including species such as the Green Darner (*Anax junius*) and the Blue Dasher (*Pachydiplax longipennis*). The increased water volume and habitat complexity in these environments allow for a greater variety of ecological niches, supporting a more diverse dragonfly community.

Preserving habitat diversity is crucial

The distribution of dragonflies and damselflies across these habitats is often patchy and highly specific, reflecting the insects' sensitivity to environmental conditions. Factors such as water chemistry, vegetation structure, and the presence or absence of certain predators can dramatically influence which species are present in a given habitat. This specificity means that careful observation of Odonata species can provide valuable information about the characteristics and health of a wetland ecosystem.

For instance, the presence of certain species can indicate the permanence of a water body, its trophic status, or even its water quality. The Azure Bluet (*Enallagma aspersum*), for example, is often associated with acidic waters, while the presence of the Elfin Skimmer (*Nannothemis bella*) can indicate a bog or fen habitat. By paying attention to which species are present in which habitats, you can gain a deeper understanding of both the habitat and its inhabitants.

This relationship between Odonata and their habitats is not unidirectional. The dragonflies and damselflies play crucial roles in these ecosystems, serving as both predators and prey. Their larvae, known as nymphs, are voracious aquatic predators, influencing the populations of other invertebrates and even small vertebrates. As adults, dragonflies and damselflies continue to be significant predators while also serving as food for birds, fish, and other organisms.

Understanding the specific Odonata communities associated with different wetland habitats enriches our appreciation of these ecosystems' complexity and interconnectedness. This highlights the importance of preserving a diverse array of wetland types to maintain biodiversity. Moreover, it underscores the value of careful observation and species identification in ecological studies. By recognizing the distinct dragonfly and damselfly communities in vernal pools, small ponds, and larger water bodies, we gain a more nuanced understanding of wetland ecosystems and the intricate relationships that sustain them.

Blue Dasher
Pachydiplax longipennis

Azure Bluet
Enallagma aspersum

Elfin Skimmer
Nannothemis bella

Common Green Darner
Anax junius

An otherwise healthy and well-maintained park pond fights a losing battle against excess nutrients that enter the water from a local suburban stream containing lawn fertilizer runoff and other pollutants common to developed landscapes. Although polluted park ponds are hardly ideal, such ponds and other wetlands are nonetheless valuable buffers to the hurricanes, rainstorms, and other violent weather that now threaten our developed landscapes every year.

THREATS TO PONDS AND WETLANDS

The Eastern United States is home to diverse freshwater ecosystems, including ponds, small lakes, marshes, and swamps. These areas are crucial to the region's ecology, hydrology, and climate regulation. However, wetlands face unprecedented challenges due to climate change and human activities that pollute water.

Climate change

Global warming has significant implications for ponds and small lakes in the Eastern United States. Due to their relatively small size and shallow depths, these water bodies are particularly susceptible to changes in temperature. These aquatic systems absorb more heat as air temperatures rise, increasing water temperatures throughout the year.

One of the most immediate impacts of warmer water is reduced dissolved oxygen levels (*see illustrations, p. 15*). Warmer water holds less dissolved oxygen, which creates challenging conditions for aquatic organisms, especially those adapted to more relaxed, oxygen-rich environments. Reductions in dissolved oxygen levels in water can lead to shifts in species composition, with cold-water species facing potential local extinctions and warm-water species possibly expanding their ranges.

Warmer temperatures also affect the annual changes in freshwater environments. Many ponds and small lakes in temperate regions undergo seasonal mixing, in which surface and deeper waters mix due to temperature-driven density differences (*see illustration, p. 16*). Climate warming can disrupt this crucial process, leading to longer periods of temperature stratification. Extended stratification can have cascading effects on nutrient cycling, oxygen distribution, and overall ecosystem functioning.

Rain gardens are designed low spots in the home landscape that capture and filter stormwater runoff from roofs, driveways, and lawns. These gardens incorporate native plants with deep root systems, permeable soil mixtures, and often a gravel or stone layer. The surface gravel and stones help the water drain into the soil and act as a physical barrier to rapid stormwater flows.

Rain gardens encourage water infiltration into the ground around the house, reducing runoff to street drains and improving water quality by trapping pollutants such as excess fertilizers. Benefits include enhanced groundwater recharge, reduced erosion, increased biodiversity in your neighborhood, and aesthetic appeal. Rain gardens are an efficient, low-maintenance solution for managing residential stormwater, contributing to local ecosystem health and sustainable water management practices.

Hurricane Helene damage to a natural mountain river near Asheville, North Carolina, October 3, 2024. The violent deluges from storms and hurricanes can severely damage natural wetland areas, but wetlands can provide very effective protection to human infrastructure if we leave wetlands and natural river corridors intact. Wetland plants and animal communities have evolved to face periodic storms and can recover more quickly and cost-effectively than most human-engineered flood controls. NCDOT photo.

The warming environment also influences the timing of seasonal events in these aquatic ecosystems. Spring events, such as ice melt and the onset of algal blooms, may occur earlier in the year. Spring amphibian breeding periods are becoming later and are often disrupted as weather patterns become more erratic. Fall events such as water cooling and leaf drop from surrounding vegetation may be delayed. These shifts can create mismatches between the life cycles of different species, potentially disrupting food webs and ecological relationships that have evolved over long periods.

Warmer conditions also often favor the growth of certain algae and aquatic plants, lead to more frequent or severe algal blooms, including potentially harmful blue-green algae (cyanobacteria). These blooms can degrade water quality, produce toxins detrimental to wildlife and humans, and create oxygen-depleted zones as they decompose, further stressing aquatic life.

Severe and frequent storms

Climate change alters precipitation patterns across the Eastern United States, leading to more severe and frequent storms. These intense weather events pose significant challenges to ponds and small lakes, disrupting their physical, chemical, and biological processes. One clear trend is that storms such as hurricanes are becoming wetter and can deliver unprecedented volumes of stormwater over short periods, as seen in the widespread destruction that followed Hurricanes Helene and Milton in 2024.

During severe storms, ponds, marshes, streams, and rivers experience rapid influxes of water, which can cause dramatic fluctuations in water levels. Sudden increases in water volume can lead to flooding, erosion of shorelines, and the resuspension of bottom sediments. These violent and sudden increases in water flow can profoundly disrupt freshwater animal and plant communities. Floodwater also severely degrades water quality in all freshwater environments. The disruption and resuspension of sediments in rivers and lakes can release nutrients and contaminants previously sequestered in the sediment. This can trigger algal blooms and bring old, long-lasting pollutants such as heavy metals and PCBs (polychlorinated biphenyls) out of bottom sediments and into the water column, severely degrading water quality.

The increased frequency of severe storms also means that these ecosystems have less time to recover between events. This can lead to chronic stress on aquatic organisms and habitats, potentially altering community structures and ecosystem functions. Species less tolerant of rapid environmental changes may decline, while more resilient or opportunistic species may thrive, leading to shifts in biodiversity.

Moreover, severe storms can physically damage aquatic and riparian vegetation. Strong winds and heavy rainfall can uproot plants, destabilize shorelines, and alter habitat structures, which can have cascading effects on the organisms that depend on these habitats for shelter, feeding, or reproduction.

In early October 2024, Hurricane Helene dumped over 30 inches of rainfall on the mountain communities of North Carolina in under 48 hours. Human infrastructure and conventional stormwater controls are proving to be insufficient in protecting roads, water supply systems, water treatment systems, homes, and other buildings. Natural wetlands can act as a buffer to protect buildings and roads from the force of stormwater. Marshall, North Carolina, October 3, 2024. NCDOT photo.

Heavy rainfall events

The increase in severe storms is closely related to the trend toward more frequent heavy rainfall. These intense precipitation episodes can profoundly impact ponds and small lakes throughout the Eastern United States.

A key challenge posed by heavy rainfall is increased runoff from surrounding landscapes. This runoff often carries elevated sediments, nutrients, and pollutants into water bodies. The influx of sediments can increase turbidity, reduce light penetration, and potentially affect primary production by aquatic plants and algae. It can also alter bottom habitats through sedimentation, potentially smothering benthic organisms and spawning areas.

The increased nutrient input associated with heavy rainfall events can lead to eutrophication, a process in which excess nutrients stimulate excessive plant and algal growth, harming water quality and aquatic life. Severe eutrophication can lead to hypoxic or anoxic conditions, creating "dead zones" where most aquatic life cannot survive.

Heavy rainfall can also cause rapid changes in water chemistry. The influx of rainwater, typically more acidic than the receiving water bodies, can lead to sudden drops in pH (acidity or alkalinity). This can stress aquatic organisms and potentially release toxic metals from sediments. The sudden dilution effect can also alter the concentration

Amphibians such as this Green Frog (*Lithobates clamitans*) live their entire life cycles in intimate contact with freshwater, and this makes them extremely vulnerable to pollutants. In addition to being harmed by obvious pollutants such as fertilizer and herbicide runoff, amphibians and other aquatic life also suffer when our pharmaceuticals enter freshwater ecosystems in household wastewater. These highly bioactive drugs can disrupt the natural development of many types of aquatic life. Here a young Green Frog has severely underdeveloped rear legs. Biologists find that such developmental and genetic defects are increasingly common in aquatic animals.

of dissolved substances, potentially disrupting aquatic life's osmotic balances.

Furthermore, these intense precipitation events can overwhelm the natural buffering capacity of wetlands and riparian zones surrounding ponds and small lakes. These areas typically act as filters, trapping sediments and nutrients before they enter the water body. However, when inundated with large volumes of water, their effectiveness can be significantly reduced, allowing more pollutants to reach the aquatic ecosystem.

Pollution from development

Pollution resulting from development in and around wetlands interacts with and often exacerbates the challenges posed by climate change. Urban development and suburban expansion in the Eastern United States frequently encroaches on or directly impacts wetland areas.

Development typically increases the amount of impervious surface area in a watershed. This includes roads, parking lots, buildings, and other structures that prevent water from infiltrating the ground. As a result, a larger volume of water flows directly into nearby water bodies, carrying a suite of pollutants during rainfall events. Typical contaminants in stormwater include fertilizers from suburban lawns, petroleum products of all kinds, lead and other heavy metals from vehicles, microplastics, and pesticides and herbicides from lawns and developed landscapes.

Introducing these pollutants into ponds and small lakes can severely affect water quality and ecosystem health. Excess nutrients can lead to eutrophication, while toxic substances can directly harm aquatic organisms. Some pollutants, like certain pesticides and heavy metals, can bioaccumulate in the food chain, potentially affecting aquatic life, terrestrial animals, and humans that depend on these ecosystems.

Moreover, development often destroys or degrades natural buffer zones around water bodies. These pondside buffer zones (*see illustration, pp. 40–41*), typically composed of natural vegetation, play crucial roles in filtering runoff, stabilizing shorelines, and providing habitat for various species. Their loss increases pollution input and reduces the resilience of these ecosystems to other stressors, including those related to climate change.

The combination of increased pollution from development and the hydrological changes brought about by climate change creates a particularly challenging scenario for ponds and small lakes. More frequent and intense rainfalls can mobilize and transport more significant quantities of pollutants into these water bodies, potentially overwhelming their natural capacity to process and mitigate these inputs.

The term "water pollution" calls to mind the old sins of industry using our wetlands as sewers, but since the passage of the federal Clean Water Act of 1972 this is much less common. Today, the more routine runoff from developed urban and suburban landscapes and street drainage is the major threat to freshwater wetlands.

Vibrant, healthy, and *numerous* wetlands are among our most important and cost-effective buffers to the effects of violent weather and climate change.

Amphibian vulnerability to pollution and climate change

The distinctive physiology and life cycle of amphibians such as frogs, toads, and salamanders makes them uniquely susceptible to water pollution and climate change. Their permeable skin controls gas exchange and osmoregulation, so they are highly sensitive to environmental contaminants. Their intimate relationship with water renders them vulnerable to pollutants in aquatic ecosystems. Chemicals, heavy metals, and other toxins can be absorbed through their skin, leading to physiological disruptions and potential reproductive failures.

Climate change compounds these challenges by altering temperature and precipitation patterns crucial for amphibians' survival. Many species rely on specific environmental cues for breeding, and shifts in these patterns can desynchronize reproductive cycles. Rising temperatures can increase the prevalence of pathogens, such as the chytrid fungus, which has decimated amphibian populations worldwide. Climate-induced habitat loss and fragmentation further stress these populations, limiting their ability to adapt or migrate.

The amphibian life cycle typically involves aquatic and terrestrial stages, exposing these animals to a broader range of environmental hazards than many other vertebrates. This dual-environment dependency makes amphibians particularly sensitive to ecological disruptions in aquatic and terrestrial habitats, serving as early indicators of ecosystem health. The decline of amphibian populations often signifies broader environmental degradation, underscoring their role as sentinel species in global conservation efforts.

Ecosystem services from freshwater environments

Despite challenges, ponds, marshes, and swamps in the Eastern United States provide crucial ecosystem services to human communities. These services become even more vital as the climate changes and storms become more frequent and severe.

Water quality improvement

Water quality improvement is one of the most significant services these freshwater ecosystems provide. Wetlands, including ponds, marshes, and swamps, act as natural filters for water flowing through the landscape. Wetlands trap sediments, absorb excess nutrients, and process various pollutants, effectively cleaning the water before it reaches larger water bodies or groundwater reserves. Aquatic plants and algae uptake nutrients such as nitrogen and phosphorus that would otherwise contribute to eutrophication in downstream water bodies. In addition, the roots of wetland plants help stabilize sediments, reducing erosion and turbidity.

Microorganisms in the soil and water of these ecosystems also contribute to water purification. They break down organic matter and transform nutrients into forms that plants can use or release harmlessly into the atmosphere. Some microorganisms can even degrade certain pollutants, including petroleum products and pesticides.

As climate change leads to more intense rainfall events and increased runoff, the water purification services of these ecosystems become even more critical. Freshwater environments can mitigate the increased pollution load associated with these events, providing a buffer against water quality degradation in the broader landscape.

Flood mitigation

Ponds, marshes, and swamps play a vital role in flood mitigation. These water bodies and the vegetation within them act as natural sponges and physical buffers against stormwater flow, absorbing and slowly releasing excess water, thereby reducing peak flood levels and flow rates. Plant stems and leaves create friction, reducing the velocity of floodwaters and giving water more time to be absorbed into the soil or evaporate. Swamps and floodplain wetlands have evolved to withstand periodic flooding. They often recover quickly from flood events and maintain their capacity to provide this crucial service even under the more frequent flooding scenarios predicted by climate change. Wetland vegetation plays a pivotal role in stabilizing shorelines and reducing erosion. The roots of aquatic and riparian plants help hold soil in place, protecting against the erosive forces of waves, currents, and floodwaters. This service becomes increasingly important as climate change leads to more frequent and intense storms, which can exacerbate erosion.

Hurricanes and rainstorms often hit poorer communities the hardest. Trailer parks and other inexpensive housing are often located on floodplains and other low-elevation areas vulnerable to flooding. Environmental justice is a growing concern as we face the effects of climate change. Flooding in St. Cloud, Florida, after Hurricane Ian in 2022.

VISITING PONDS

Early to mid-spring can be a great time to observe pond life since many frogs, salamanders, and snakes mate at this time and are easy to find. For example, the American Toad and the Wood Frog are generally shy creatures that are inactive and hidden during much of the daytime. But in many early spring ponds, mating balls of American Toads are conspicuous throughout the day as multiple males try to mate with single females floating at the pond surface. You'll have to look more closely to spot the very common but less conspicuous Northern Two-Lined Salamander adults: they are usually submerged as they mate and lay their gelatinous egg masses in vernal pools and streams. Wood Frogs favor the shallower areas of ponds or vernal pools in swampy wetlands and are often easy to find because they call constantly.

Mating Northern Watersnakes (*see illustration, p. 77*) are often easy to find as pairs entwined in the foliage around pond edges. The Northern Watersnake is nonvenomous, but its boldly patterned brick-red skin sometimes causes it to be mistaken for the Northern Copperhead (*see illustrations, p. 120*). The Northern Copperhead is a venomous species that is common in some areas and generally lives on rocky upland slopes and in dry meadows, not usually along pond edges. The chances of encountering a venomous snake in most wetland are low, but not zero. Always respect wild snakes, use great caution especially if you have pet dogs with you, and never attempt to harm or move snakes.

The Northern Watersnake (*Nerodia sipedon*) is a semiaquatic snake found in North America, including parts of the Northeastern United States. It is a common species often encountered near bodies of water, such as rivers, streams, and lakes. Northern Watersnakes are heavy-bodied snakes with an average length of two to four feet. They have a distinct pattern of large, dark brown or black patches on a lighter brown or gray background. Northern Watersnakes are gener-

Move too quickly or make too much noise and you'll miss much of the wildlife of the pond. Aquatic wildlife like this American Bullfrog have excellent vision and hearing and are also very sensitive to the vibration of careless footfalls.

ally nonaggressive and usually retreat into the water when threatened. However, they will defend themselves if cornered and can deliver a painful but nonvenomous bite.

In summer, most frog species are less visually conspicuous, but the common Green Frogs and American Bullfrogs (*see illustrations, pp. 114–15*) are easy to spot if you are patient and quiet. When you slowly approach the pond edge, use your binoculars to scan the pond's edges, just above where the pond foliage meets the water. Adult frogs like to perch under the overhanging leaves along the pond edge, where they can more easily spot prey and jump into the water to escape predators such as herons. In ponds with water lilies, scan the surface near the lily flowers and leaves for the heads of frogs poking up above the surface. American Bullfrogs frequently bask in the warm surface water of ponds, using water lily leaves as shelter and hiding places. Common dragonfly species such as the Blue Dasher and the Slaty Skimmer (*see illustrations, pp. 102–3*) often use floating water lily leaves as perches.

Always carefully scan pond and lake edges for larger birds, such as the ubiquitous Great Blue Heron and the bright-white Great Egret (*see illustrations, pp. 246–47*). The black-billed Snowy Egret is less common and resembles a smaller Great Egret. The Black-Crowned Night-Heron is another common, if stealthy, large avian predator in freshwater wetlands. Night-herons are relatively short-legged; as their name suggests, they are most active between dusk and dawn. During the day,

This schedule of frog mating calls gives an approximate time range for each species, but many frog species call only for a few weeks each year. The exact timing for spring calling depends on local weather and geography. The temperature in colder bottomland areas or sheltered valleys may cause frogs to call much later, and in warmer areas the frogs may call much earlier in their "scheduled" time window. It pays big dividends to constantly check your favorite wetlands so that you don't miss a mating event.

Schedule of Frog Calls in Spring and Summer

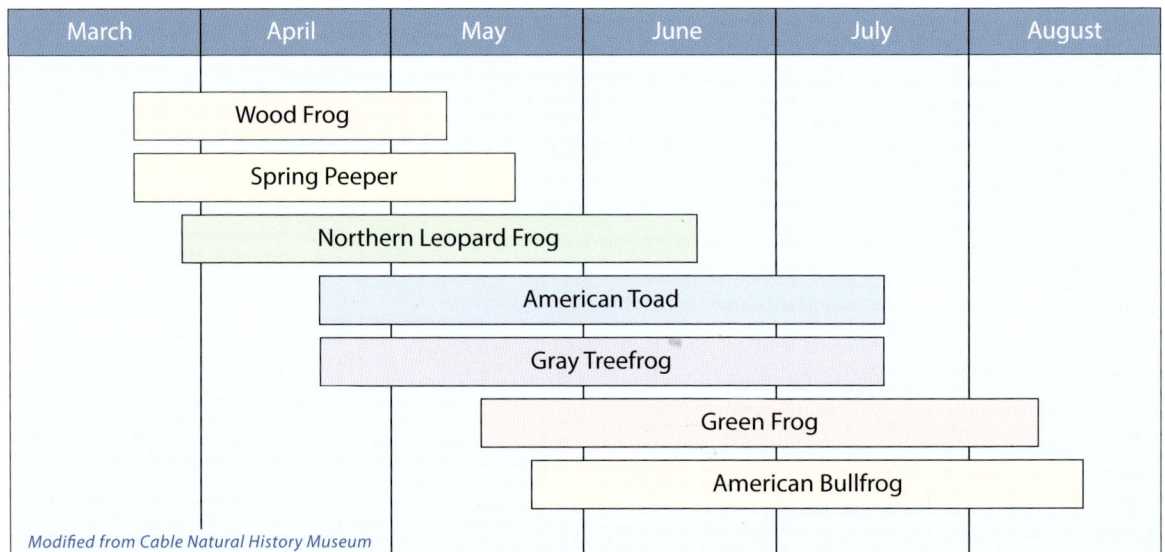

March	April	May	June	July	August
	Wood Frog				
	Spring Peeper				
	Northern Leopard Frog				
		American Toad			
		Gray Treefrog			
			Green Frog		
			American Bullfrog		

Modified from Cable Natural History Museum

night-herons tend to perch on branches above or very near water, so it is always worthwhile to scan pondside trees for roosting night-herons. As you move along lake shores at dusk, you will sometimes hear the loud "kwok!" of a night-heron driven off its perch by your approach.

From mighty regional rivers down to the smallest neighborhood streams and ponds, there's no waterbird more conspicuous or characteristic than the Belted Kingfisher (*see illustrations, pp. 55, 84*). The kingfisher's loud, rattling call usually announces its presence near any water containing small fish. The Belted Kingfisher's outsized head and bill make it look larger than it is: the bird is only slightly larger than a Blue Jay. Kingfishers are one of the few bird species that can truly hover in place, which they do to spot small fish beneath the pond or stream surface. They then fold their wings and make a headfirst dive into the water, usually emerging with a fish in their beak.

Although you might not think of owls as waterbirds, the Eastern Screech Owl is an aggressive hunter in all freshwater environments. These owls will enter the shallow water of ponds and streams in search of larger insects, frogs, crayfish, and whatever else they can capture. They are more common than most birders realize and are seldom found far from wetlands. If you want to see (or, more likely, hear) your first Eastern Screech Owl, quietly hang out at a favorite pond or stream as dusk turns to true darkness about half an hour after sunset. The owl's high, wavering call is very distinctive at close range. But

A mating pair of Northern Watersnakes beside a small park pond in suburban Connecticut. Watersnakes are quite common in wetlands. With their sometimes dramatic red patterning, the nonvenomous Northern Watersnakes are often mistaken for the less common but venomous Northern Copperhead. See the illustration on pp. 120 to learn the differences between the two species. ***Please*** never try to harm or move a wild snake of any species. Snakes of all kinds are vital to a balanced ecosystem.

when filtered through trees with a breeze blowing, the call is sometimes hard to distinguish from a distant dog howl or a wailing siren. Wetlands also attract larger owls, including the Barred and Great Horned Owls. If you hear them calling, appreciate your good luck, but give up on hearing screech owls that night. Barred Owls and Great Horned Owls will readily hunt the much smaller Eastern Screech Owl, and their calls will typically silence the smaller raptor.

Observing pond wildlife

You should always slowly approach pond or lake edges, avoiding sudden noises or talking with companions. Often, the first indication that you might have moved too quickly is the "plop" sounds of frogs hopping from their pondside perches and splashing into the water. Vertebrates such as fish, frogs, and turtles seem to recognize you as a fellow animal and potential predator, so approach the pond carefully and quietly and scan the area thoroughly before breaking cover and becoming more visible to pond animals. However, large insects such as dragonflies, various water bugs, and butterflies respond only to sudden movement and don't seem to distinguish humans from the rest of the pond environment. If you are still enough for long enough, it's not uncommon to have dragonflies perch on your binoculars or camera.

EASTERN SCREECH OWL
Megascops asio

BLACK-CROWNED NIGHT-HERON
Nycticorax nycticorax

YELLOW-CROWNED NIGHT-HERON
Nyctanassa violacea

GREEN HERON
Butorides virescens

The very stealthy Green Heron is the most common avian predator in most freshwater ponds, aside from the noisy, conspicuous, and even more common Belted Kingfisher. Both the Black-Crowned and Yellow-Crowned Night-Herons can also be found in freshwater pond ecosystems, although the Black-Crowned is more common. The Yellow-Crowned Night-Heron is more frequently seen in coastal plain freshwater and brackish ponds.

BLUE DASHER (male)
Pachydiplax longipennis

Many dragonflies, such as the ubiquitous Blue Dasher, are perch hunters and habitually return to their favorite prominent branches that give them a good field of view. For closer views or better photographs, watch the overall behavior pattern of pond dragonflies and then move closer to a favorite perch. You'll probably scare away the dragonflies at first, but if you are patient and still, they'll return very quickly.

Chasing active fliers such as butterflies and dragonflies is pointless. Your sudden movements will only drive the animals farther away, and all the thrashing and noise will scare away any nearby animals. Patient waiting is a much better strategy, as whatever nearby flower or perch attracted the fliers in the first place will certainly attract other dragonflies or butterflies.

Wading out into a pond or lake edge to observe smaller aquatic animals isn't practical for most wetland visitors. Traditional observation and collection can disrupt heavily visited park wetland environments. But there are techniques to help bridge the awkward gap between the pond and the shore. Most hikers and avid naturalists are also bird watchers and own a pair of binoculars. Birds are the most popular form of wildlife because they are both easily visible and diurnal—that is, they are active during daylight hours. But binoculars are also perfect tools to observe many forms of small-to-medium-sized wildlife on the edges of ponds.

Many current roof prism 8 × 42 (or similar) binocular models can focus as close as four to six feet and are perfect for adding dragon-

The most popular binoculars for wildlife watching are straight-tube roof-prism designs. Midpriced or premium-priced roof-prism designs offer sharp optics, reasonable brightness, and much better close focusing.

fly, butterfly, and aquatic animal observation to your bird-watching routines. If you are buying a pair of binoculars for general use, go to a local birding, wildlife, or photography store and get expert advice. Test a range of models for brightness and viewing comfort and pay particular attention to each model's minimum focusing distance.

Large insect observation complements birding, as by midmorning on sunny summer days, most bird species have become less active but butterflies, damselflies, dragonflies, and bees are just beginning their period of highest activity. Hence, as the midday bird activity fades, you can quickly shift to observing other pond wildlife. Butterflies are attracted to various plants along pond edges and wet meadow areas and may also concentrate on exposed wet mud or damp pond banks. Carefully scan any flowers present, particularly for smaller species, including skippers and bumblebees. On breezy days, try to favor the downwind shore of the pond, as the wind will carry butterflies and dragonflies across the water and concentrate them on the leeward shore. Many dragonfly species are perch hunters that habitually return to their favorite twigs or branches near the pond edges. Be patient and still. If you scare a dragonfly away from its perch, it will almost certainly return, sometimes within seconds.

Avoid these older Porro-prism designs. While some older Porro models from reputable brands such as Nikon are of good quality, cheaper Porro-prism designs will be dimmer and less sharp than current roof-prism designs. Older Porro-prism binoculars also rarely focus as close as current designs, a big factor if you are interested in watching butterflies and dragonflies.

MONARCH BUTTERFLY
Danaus plexippus

Staying safe and comfortable

Awareness and careful preparation are the keys to enjoying wild environments safely. The most severe threats most people will face outdoors are from biting insects and ticks, not from bears, coyotes, snakes, or other larger animals.

Clothing for the outdoors

Shorts and a T-shirt are fine for a stroll on wide park trails on a breezy day. But any time you are likely to find yourself brushing against tall grasses or tree branches, you should be dressed in long pants and a long-sleeved shirt. Long clothing can be daunting on a hot summer day, so shop carefully for shirts and pants made from light synthetic cloth that won't be too uncomfortable in the heat. Warm-weather fly fishing clothing is often ideal for hiking and exploring wetlands.

Ideally, your field clothing should be treated with permethrin, as this insecticide for clothing is the only thing that will reliably discourage or kill the ticks that spread Lyme disease, babesiosis, and other serious tick-borne illnesses. You can buy treated clothing off-the-shelf at places specializing in hiking, hunting, and other outdoor sports. However, the effectiveness of pretreated clothing fades with multiple washings, and you'll want to treat the clothing yourself with permethrin spray. **It's important to understand that permethrin is never used directly to treat your skin.** Permethrin is only used to treat clothing. Always carefully follow the package directions for treating clothing.

Wetlands can be buggy places, particularly when there is little breeze to hold down flying insects. Most people who spend time outdoors use DEET-based repellent sprays to ward off mosquitoes. DEET sprays are considered safe and effective when used as directed. Non-DEET insect repellent sprays may have various advantages, but reliably preventing mosquito bites isn't one of them. DEET is a chemical used for decades in insect repellents and approved by the Environmental Protection Agency (EPA). Studies have shown it is effective at repelling mosquitoes, discouraging ticks, and biting flies. However, there are some potential risks associated with DEET use, such as skin irritation or allergic reactions in some people. It is essential to follow the label instructions carefully and avoid using DEET on young children or broken skin. Generally, it's best to use a DEET product with the lowest concentration that provides adequate protection for your activities and environment.

The right footwear

Wetlands are wet, and it's almost impossible not to get regular shoes soaked when exploring ponds, marshes, and swamps. You could wear old shoes that you don't mind getting wet and muddy, but wet shoes and socks feel lousy, and you'll risk blisters or chilly feet on a longer hike. Invest in a sturdy and comfortable pair of waterproof boots if

Lone Star Tick
Amblyomma americanum

Blacklegged Tick
(Deer Tick)
Ixodes scapularis

American Dog Tick
Dermacentor variabilis

Freshwater environments harbor several species of ticks that mostly parasitize wild mammals such as White-Footed Deer Mice, Meadow Voles, Raccoons, White-Tailed Deer, and Red Foxes. Always wear long pants when you explore away from trails in marshes and pond-side vegetation. Apply a DEET-based or Permethrin insect repellent on your clothes, socks, and shoes, as directed by the manufacturer.

you regularly visit wetland areas for birding or nature study. Ideally, the boots should be tall enough to cover your calves and have tough, nonslip soles. This kind of boot also gives your lower legs excellent protection from ticks and bug bites. Waterproof outdoor boots are standard equipment for hunters, so stores or online sites specializing in hunting clothing can be great places to find suitable boots and effective treated outdoor clothing.

Know where you are

Most parks and recreational areas can provide trail maps, so always look online for information before you head into the wilds. Your smartphone can be an excellent tool for keeping yourself oriented to your location, but make sure that your phone has sufficient cell signal to be useful. Hills and trees can hinder cell phone signals; you may get no signal in more remote places. Do your research and carry a paper trail map and a basic trail compass in your backpack or camera bag.

BELTED KINGFISHER (female)
Megaceryle alcyon

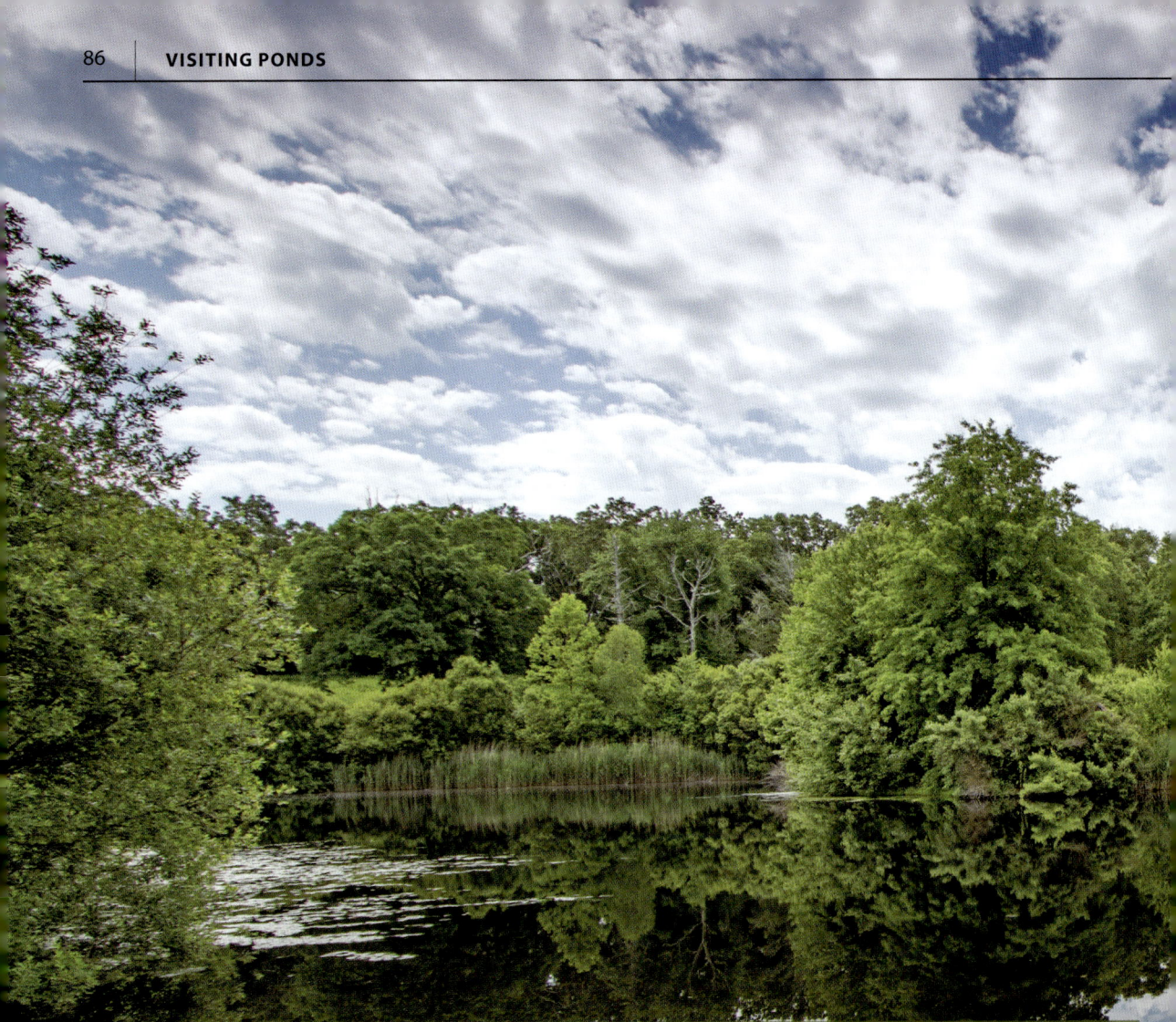

A well-designed and maintained state park pond in central Connecticut. Only in one small area (where the photographer was standing) does the park lawn approach the pond edge. Importantly, the park lawns are not artificially fertilized, to prevent lawn runoff from polluting the park's ponds. Note the native Black Willow, Smooth Alder, Buttonbush, and other shrubs around the pond edges. This protective wall of shrubs shields much of the pond from casual human visitors who might trample the pond-edge vegetation and thereby muddy the pond water. Multiple small gaps in the shrub wall allow fishers to reach the pond. The shrub wall also acts to catch, filter, and absorb runoff from the surrounding park lawns. The result: wonderfully clear pond water, free of excess algae, and an excellent home for the wildlife of the pond.

Machimoodus State Park, Moodus, Connecticut.

COMMON FRESHWATER PLANTS AND ANIMALS

An introductory book such as this can't cover all of the thousands of plant and animal species in freshwater wetlands. For deeper dives into particular groups, including wildflowers, butterflies, and birds, see "Further Reading."

Great Blue Heron
Ardea herodias

See also "Northern Bogs and Southern Pocosins."

SKUNK CABBAGE *Symplocarpus foetidus*

PICKERELWEED *Pontederia cordata*

UMBRELLA SEDGE *Cyperus alternifol*

ARROWHEAD *Sagittaria latifolia*

ARROW ARUM *Peltandra virginica*

TUSSOCK SEDGE *Carex stricta*

COMMON CATTAIL *Typha latifolia*

NARROW-LEAVED CATTAIL *Typha angustifolia*

NORTHERN WILD RICE *Zizania palus*

See also "Freshwater Marshes."

ERICAN BUR-REED *Sparganium americanum*

WOOLGRASS *Scirpus cyperinus*

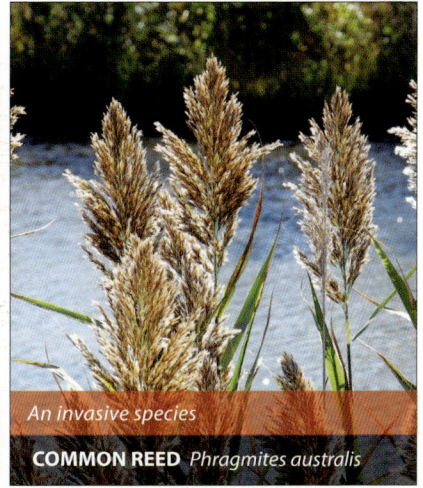

An invasive species

COMMON REED *Phragmites australis*

NAMON FERN *Osmundastrum cinnamomeum*

SENSITIVE FERN *Onoclea sensibilis*

OSTRICH FERN *Matteuccia struthiopteris*

YAL FERN *Osmunda regalis*

MARSH FERN *Thelypteris palustris*

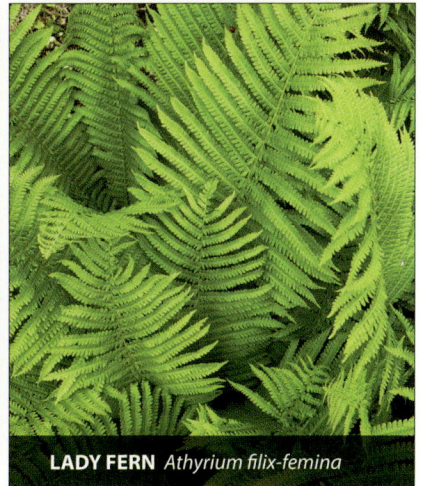

LADY FERN *Athyrium filix-femina*

Many of these species are not strictly wetland plants but commonly occur near and on the edges of wetlands.

BLOODROOT *Sanguinaria canadensis*

OXEYE DAISY *Leucanthemum vulgare*

BONESET *Eupatorium perfoliatum*

JACK-IN-THE-PULPIT *Arisaema triphyllum*

TROUT LILY *Erythronium americanum*

BLUETS *Houstonia caerulea*

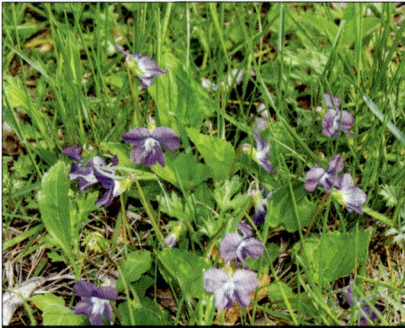

COMMON BLUE VIOLET *Viola sororia*

RED TRILLIUM *Trillium erectum*

PAINTED TRILLIUM *Trillium undulatu*

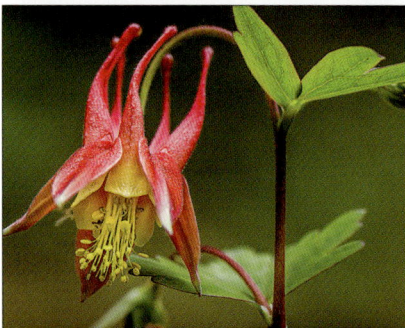

EASTERN RED COLUMBINE *Aquilegia canadensis*

MARSH MARIGOLD *Caltha palustris*

BUTTERCUP *Ranunculus bulbosus*

OTTED JOE-PYE WEED *E. maculatum*

CARDINAL FLOWER *Lobelia cardinalis*

GOLDENROD *Solidago sp.*

RPLE LOOSESTRIFE *Lythrum salicaria*

vasive species

HARLEQUIN BLUEFLAG *Iris versicolor*

YELLOW FLAG *Iris pseudacorus*

R-MARIGOLD *Bidens cernua*

PICKERELWEED (flower) *Pontederia cordata*

ARROWHEAD (flower) *Sagittaria latifolia*

WELWEED *Impatiens capensis*

WHITE WATER-LILY *Nymphaea alba*

YELLOW WATER-LILY *Nuphar lutea*

Many of these species are not strictly wetland plants but commonly occur near and on the edges of wetlands.

SWAMP MILKWEED *Asclepias incarnata*

COMMON MILKWEED *Asclepias syriaca*

BUTTERFLY WEED *Asclepias tuberosa*

ORANGE HAWKWEED *Hieracium aurantiacum*

RAGGED ROBIN *Silene flos-cuculi*

FIREWEED *Chamaenerion sp.*

WILD GERANIUM *Geranium maculatum*

DAME'S ROCKET *Hesperis matronalis*

SWAMP ROSE *Rosa palustris*

DEPTFORD PINK *Dianthus armeria*

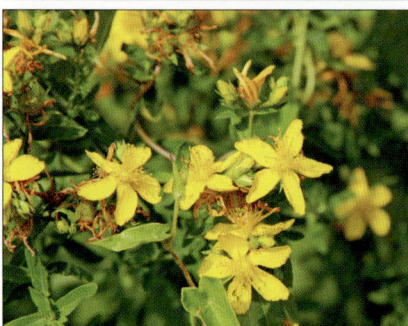

ST. JOHN'S WORT *Hypericum perforatum*

NEW ENGLAND ASTER *Symphyotrichur*

JE-PYE WEED *Eutrochium purpureum*

SWAMP ROSE MALLOW *Hibiscus moscheutos*

FIELD BINDWEED *Convolvulus arvensis*

ATE PURPLE ASTER *Symphyotrichum patens*

SWAMP AZALEA *Rhododendron viscosum*

PINK AZALEA *Rhododendron periclymenoides*

OONTAIL *Ceratophyllum demersum*

WATERWEED (ELODEA) *Elodea sp.*

WATERMILFOIL *Myriophyllum sp.*

JRFACE SPIROGYRA *Spirogyra sp.*

COMMON DUCKWEED *Lemna minor*

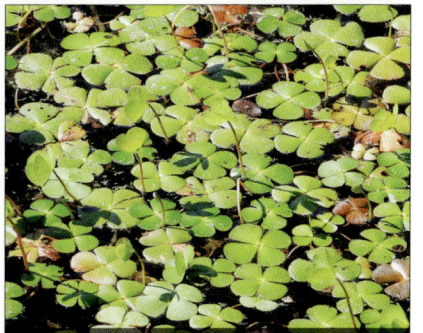

WATER CLOVER *Marsilea quadrifolia*

See also "Freshwater Swamps and Shrub Swamps."

BUTTONBUSH *Cephalanthus occidentalis*

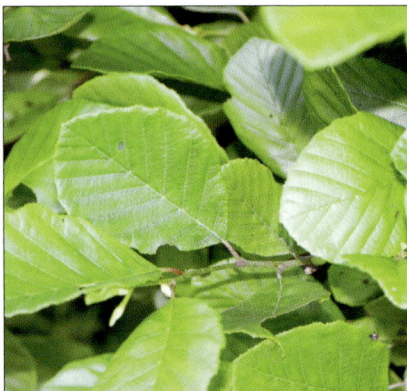

There are three very similar alder species

SMOOTH ALDER *Alnus serrulata*

HIGHBUSH BLUEBERRY *Vaccinium s*

BLACK WILLOW *Salix nigra*

Poisonous to touch!

POISON IVY *Toxicodendron radicans*

Poisonous to touch!

POISON SUMAC *Toxicodendron vernix*

MOUNTAIN LAUREL *Kalmia latifolia*

NORTHERN SPICEBUSH *Lindera benzoin*

REDOSIER DOGWOOD *Cornus sericea*

MOOTH SUMAC *Rhus glabra*

STAGHORN SUMAC *Rhus typhina*

SWEET PEPPERBUSH *Clethra alnifolia*

HINING SUMAC *Rhus copallinum*

AUTUMN OLIVE *Elaeagnus umbellata*

MULTIFLORA ROSE *Rosa multiflora*

RROWWOOD VIBURNUM *V. dentatum*

INKBERRY *Ilex glabra*

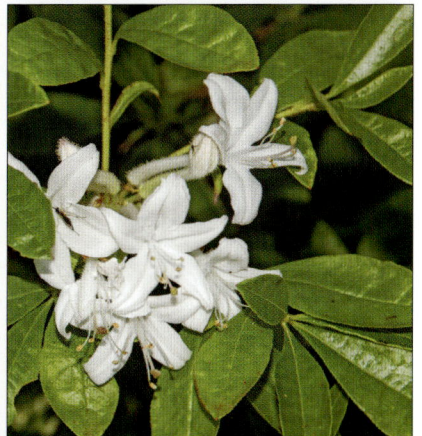

SWAMP AZALEA *Rhododendron viscosum*

See also "Freshwater Swamps and Shrub Swamps."

EASTERN COTTONWOOD *Populus deltoides*

QUAKING ASPEN *Populus tremuloides*

RED MAPLE *Acer rubrum*

BLACK TUPELO *Nyssa sylvatica*

PAPER BIRCH *Betula papyrifera*

YELLOW BIRCH *Betula alleghaniensis*

AMERICAN HORNBEAM *Carpinus caroliniana*

BLACK ASH *Fraxinus nigra*

PUSSY WILLOW *Salix discolor*

VER MAPLE *Acer saccharinum*

AMERICAN SYCAMORE *Platanus occidentalis*

BOXELDER MAPLE *Acer negundo*

ACK CHERRY *Prunus serotina*

AMERICAN HOLLY *Ilex opaca*

SWEETGUM *Liquidambar styraciflua*

VE OAK *Quercus virginiana*

LOBLOLLY PINE *Pinus taeda*

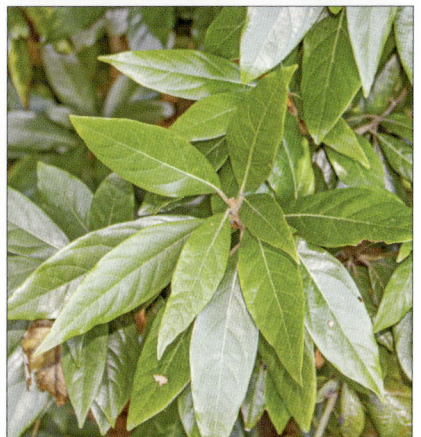

REDBAY *Persea borbonia*

Note: Small invertebrates often require professional expertise to identify to the species level.

WATER STRIDER *family Gerridae*

WHIRLIGIG BEETLES *family Gyrinidae*

BACKSWIMMER *family Notonectidae*

WATER BOATMAN *family Corixidae*

GIANT WATER BUG *Lethocerus americanus*

DIVING BEETLE *family Dytiscidae*

POND LEECH *family Hirudinea*

FISHING SPIDER *Dolomedes sp.*

CRAYFISH *Various families, about nine sp*

DRAGONFLY LARVA *order Odonata*

FRESHWATER MUSSEL *family Unionidae*

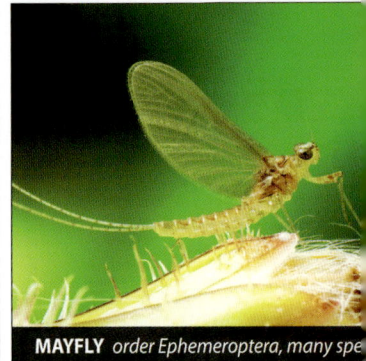

MAYFLY *order Ephemeroptera, many spe*

ONY JEWELWING *Calopteryx maculata*

FAMILIAR BLUET *Enallagma civile*

NORTHERN SPREADWING *Lestes disjunctus*

TERN FORKTAIL *Ischnura verticalis*

EASTERN RED DAMSEL *Amphiagrion saucium*

RIVER JEWELWING *Calopteryx aequabilis*

NY-EDGED SKIPPER *Polites themistocles*

LONG DASH SKIPPER *Polites mystic*

PECK'S SKIPPER *Polites peckius*

AD-WINGED SKIPPER *Poanes viator*

SILVER-SPOTTED SKIPPER *Epargyreus clarus*

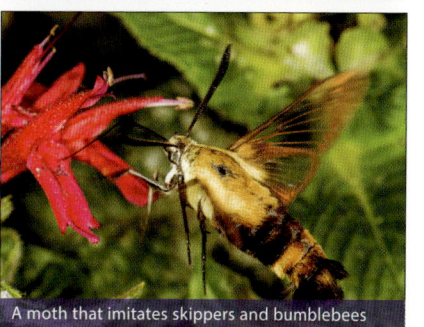

A moth that imitates skippers and bumblebees

SNOWBERRY CLEARWING *Hemaris diffinis*

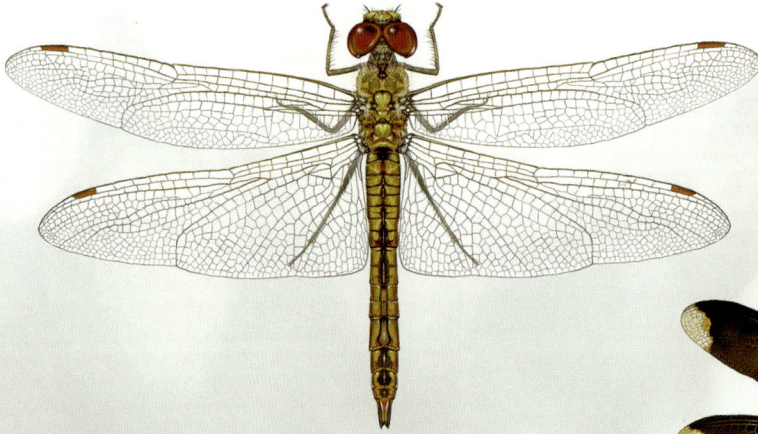

WANDERING GLIDER
Pantala flavescens

COMMON WHITETAIL
Plathemis lydia

HALLOWEEN PENNANT
Celithemis eponina

TWELVE-SPOTTED SKIMMER
Libellula pulchella

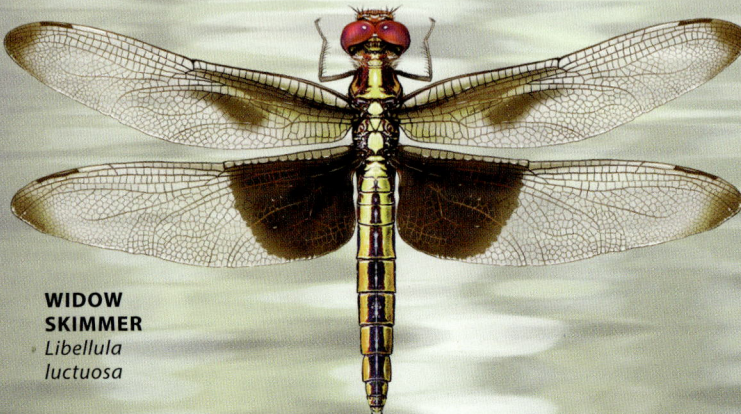

WIDOW SKIMMER
Libellula luctuosa

RUBY MEADOWHAWK
Sympetrum rubicundulum

All species are shown to scale

GREEN DARNER
Anax junius

BLACK SADDLEBAGS
Tramea lacerata

EASTERN AMBERWING
Perithemis tenera

BLUE DASHER
Pachydiplax longipennis, female

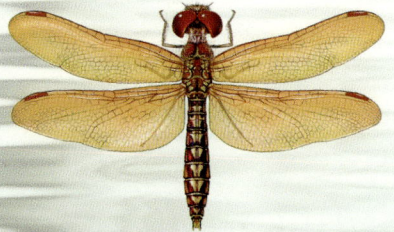

EASTERN PONDHAWK
Erythemis simplicicollis, female

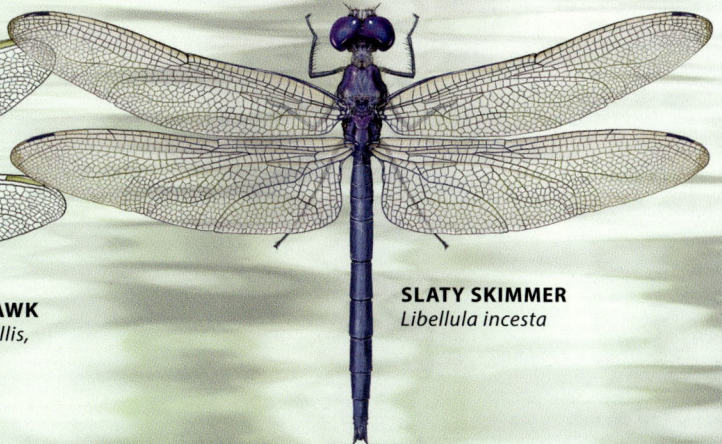

SLATY SKIMMER
Libellula incesta

All species are shown to scale

MOURNING CLOAK
Nymphalis antiopa

RED ADMIRAL
Vanessa atalanta

QUESTION MARK
Polygonia interrogationis

GRAY HAIRSTREAK
Strymon melinus

VARIEGATED FRITILLARY
Euptoieta claudia

RED-SPOTTED PURPLE
Limenitis arthemis

**EASTERN
TIGER SWALLOWTAIL**
Papilio glaucus

**SPICEBUSH
SWALLOWTAIL**
Papilio troilus

**BLACK
SWALLOWTAIL**
Papilio polyxenes

All species are shown to scale

COMMON WOOD NYMPH
Cercyonis pegala

PAINTED LADY
Vanessa cardui

COMMON BUCKEYE
Junonia coenia

NORTHERN PEARLY-EYE
Enodia anthedon

CLOUDED SULPHUR
Colias philodice

GREAT SPANGLED FRITILLARY
Speyeria cybele

VICEROY
Limenitis archippus

MONARCH
Danaus plexippus

ACADIAN HAIRSTREAK *Satyrium acadica*

BANDED HAIRSTREAK *Satyrium calanus*

CORAL HAIRSTREAK *Satyrium titus*

GRAY HAIRSTREAK *Strymon melinus*

JUNIPER HAIRSTREAK *Callophrys gryneus*

EASTERN TAILED-BLUE *Cupido com*

LITTLE WOOD-SATYR *Megisto cymela*

CABBAGE WHITE *Pieris rapae*

PEARL CRESCENT *Phyciodes tharos*

COMMON RINGLET *Coenonympha tullia*

AMERICAN COPPER *Lycaena phlaeas*

CLOUDED SULPHUR *Colias philodi*

COMMON EASTERN BUMBLEBEE *Bombus sp.*

EASTERN CARPENTER BEE *Xylocopa sp.*

COMMON LONG-HORNED BEE *Melissodes sp.*

GREAT GOLDEN SAND DIGGER *Sphex sp.*

HONEY BEE *Apis mellifera*

BICOLORED SWEAT BEE *Agapostemon sp.*

FURROW BEE *Halictus sp.*

Flower flies imitate bees

MARGINED CALLIGRAPHER FLY *Toxomerus sp.*

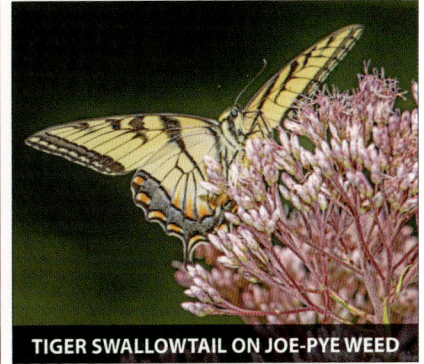

TIGER SWALLOWTAIL ON JOE-PYE WEED

FOUR-BANDED LONGHORN BEETLE

Spiders are Arachnids, not insects

BRILLIANT JUMPING SPIDER *Phidippus sp.*

TWELVE-SPOTTED SKIMMER *Libellula sp.*

BLUEGILL
Lepomis macrochirus
4–8 in.

PUMPKINSEED
Lepomis gibbosus
3–6 in.

**REDBREAST
SUNFISH**
Lepomis auritus
4–6 in.

BLACK CRAPPIE
Pomoxis nigromaculatus
6–11 in.

ROCK BASS
Ambloplites rupestris
5–7 in.

BROWN BULLHEAD
Ameiurus nebulosus
8–13 in.

WHITE CRAPPIE
Pomoxis annularis
6–11 in.

YELLOW PERCH
Perca flavescens
6–11 in.

GOLDEN SHINER
Notemigonus crysoleucas
3–6 in.

SMALLMOUTH BASS
Micropterus dolomieu
6–16 in.

WALLEYE
Sander vitreus
10–20 in.

CHAIN PICKEREL
Esox niger
10–18 in.

NORTHERN PIKE
Esox lucius
18–30 in.

COMMON CARP
Cyprinus carpio
16–28 in.

LARGEMOUTH BASS
Micropterus salmoides
6–18 in.

REDFIN PICKEREL
Esox americanus americanus
5–9 in.

WHITE SUCKER
Catostomus commersonii
6–18 in.

ALEWIFE
Alosa pseudoharengus
9–11 in.

LAKE TROUT
Salvelinus namaycush
24–36 in.

Frogs and Toads of New England
Species photos courtesy of Twan Leenders

American Toad
Anaxyrus americanus

Fowler's Toad
Anaxyrus fowleri

American Toad mating cluster
Two mating toads in amplexus are surrounded by other males competing to mate. Note the speckled linear toad egg masses below the group.

Toad mating ball photo by Pat Lync

Wood Frog tadpole
Rana sylvatica (Not to scale with the adult frogs.)

Spring Peeper
Pseudacris crucifer

Pickerel Frog
Lithobates palustris

Gray Treefrog
Hyla versicolor

Northern Leopard Frog
Lithobates pipiens

Wood Frog
Lithobates sylvatica

Species are shown approximately to scale

Bullfrog
Lithobates catesbeiana
Female–Note the white throat; males usually have bright yellow throats

Green Frog
Lithobates clamitans

Salamanders of Eastern North America
Species photos courtesy of Twan Leenders

*Species are shown
approximately to scale*

Northern Two-Lined Salamander
Eurycea bislineata 3–4.5 in. long

Female

Male

Red-Spotted Newt, adult forms
Notophthalmus viridescens 3–5 in. lon

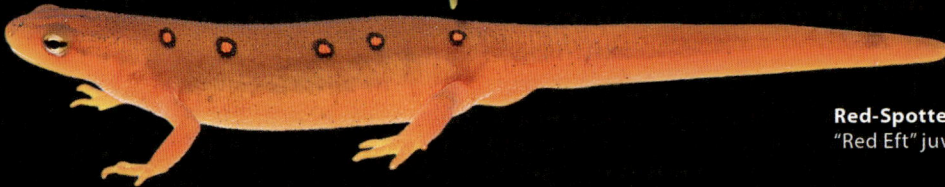

Red-Spotted Newt
"Red Eft" juvenile form

Mudpuppy
Necturus maculosus
8–17 in. long

Eastern Red-backed Salamander
Plethodon cinereus 2–4 in. long

Marbled Salamander
Ambystoma opacum 3–4.75 in. long

Spotted Salamander
Ambystoma maculatum 4.75–6.75 in. long

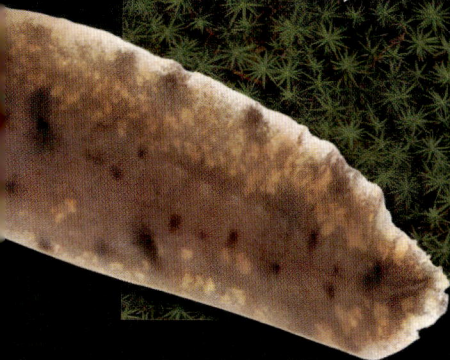

Turtles of New England
Species photos courtesy of Twan Leenders

Species are shown approximately to scale

Common Musk Turtle
Sternotherus odoratus 3–5 in. long

Spotted Turtle
Clemmys guttata 4–5 in. long

Eastern Box Turtle
Terrapene carolina 4.5–6.5 in. long

Northern Diamondback Terrapin
Malaclemys terrapin
6–9 in. long (females)

Northern Diamondback Terrapins live in the salt and brackish marshes of the East Coast.

**Eastern
Painted Turtle**
Chrysemys picta
4.5–6 in. long

Red-Eared Slider
Trachemys scripta
6–8 in. long
The Red-Eared Slider is an
invasive, non-native species on
the East Coast. Sliders that
escaped from the pet trade now
breed in many areas beyond
their original range.

Wood Turtle
Glyptemys insculpta
6–9 in. long

Common Snapping Turtle
Chelydra serpentine 10–16 in. long

NORTHERN WATERSNAKE
Nerodia sipedon
24–42 in.

Northern Watersnakes are largely harmless but are often mistaken for the venomous Northern Copperhead (see p. 77)

VENOMOUS
NORTHERN COPPERHEAD
Agkistrodon contortrix mokasen
24–47 in., average 35 in.

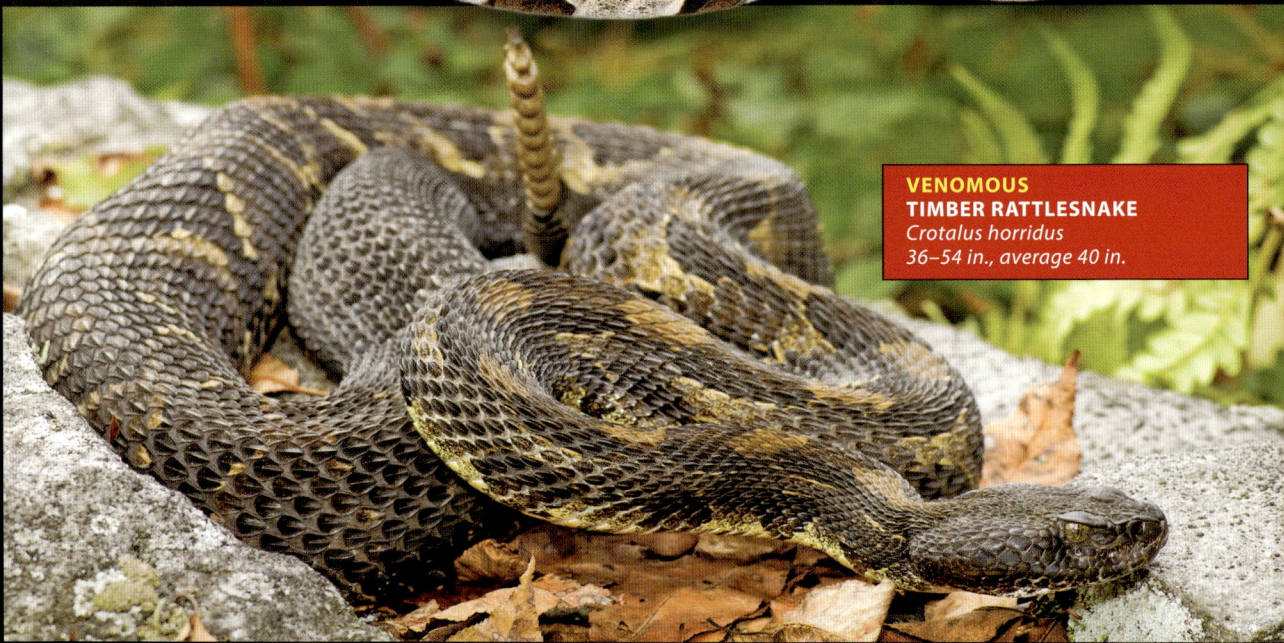

VENOMOUS
TIMBER RATTLESNAKE
Crotalus horridus
36–54 in., average 40 in.

EASTERN HOGNOSE SNAKE
Heterodon platirhinos
21–32 in.

EASTERN MILK SNAKE
Lampropeltis triangulum
19–40 in.

EASTERN RACER
Coluber constrictor
33–65 in.

EASTERN RATSNAKE
Pantherophis alleghaniensis
46–68 in.

COMMON GARTER SNAKE
Thamnophis sirtalis
18–26 in.

These smaller snakes are entirely
harmless, and all are valuable
components of forest and
wetland environments

COMMON RIBBON SNAKE
Thamnophis saurita
20–32 in.

NORTHERN REDBELLY SNAKE
Storeria occipitomaculata
8–11 in.

DEKAY'S BROWNSNAKE
Storeria dekayi
9–15 in.

See also illustrations on p. 79 for night-herons and pp. 246–247 for typical wetland herons and egrets.

TED KINGFISHER *Megaceryle alcyon*

BLACK-CROWNED NIGHT-HERON

AMERICAN BITTERN *Botaurus lentiginosus*

EAT BLUE HERON *Ardea herodias*

GREEN HERON *Butorides virescens*

GREATER YELLOWLEGS *Tringa melanoleuca*

MMON LOON *Gavia immer*

PIED-BILLED GREBE *Podilymbus podiceps*

OSPREY *Pandion haliaetus*

RED-TAILED HAWK *Buteo jamaicensis*

RED-SHOULDERED HAWK *Buteo lineatus*

OSPREY *Pandion haliaetus*

BALD EAGLE *Haliaeetus leucocephalus*

TURKEY VULTURE *Cathartes aura*

BLACK VULTURE *Coragyps atratus*

GREAT BLUE HERON *Ardea herodias*

NORTHERN HARRIER *Circus hudsonius*

GREAT EGRET *Ardea alba*

LD TURKEY *Meleagris gallopavo*

BARRED OWL *Strix varia*

KILLDEER *Charadrius vociferus*

OODCOCK *Scolopax minor*

WILSON'S SNIPE *Gallinago delicata*

COMMON GALLINULE *Gallinula galeata*

ICOLORED HERON *Egretta tricolor*

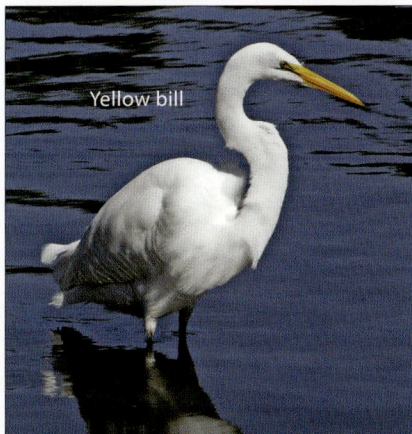

Yellow bill

GREAT EGRET *Ardea alba*

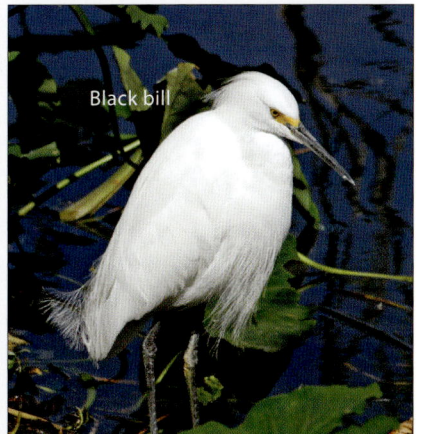

Black bill

SNOWY EGRET *Egretta thula*

CANADA GOOSE *Branta canadensis*

MALLARD *Anas platyrhynchos*

NORTHERN PINTAIL *Anas acuta*

BLUE-WINGED TEAL *Anas discors*

GREEN-WINGED TEAL *Anas carolinensis*

WOOD DUCK *Aix sponsa*

BUFFLEHEAD *Bucephala albeola*

SORA *Porzana carolina*

VIRGINIA RAIL *Rallus limicola*

ERICAN COOT *Fulica americana*

HOODED MERGANSER *Lophodytes cucullatus*

COMMON MERGANSER *Mergus merganser*

CK DUCK *Anas rubripes*

AMERICAN WIGEON *Mareca americana*

LESSER SCAUP *Aythya affinis*

WALL *Mareca strepera*

MOTTLED DUCK *Anas fulvigula*

SNOW GOOSE *Anser caerulescens*

RED-WINGED BLACKBIRD *Agelaius phoeniceus*

YELLOW WARBLER *Setophaga petechia*

COMMON YELLOWTHROAT

SONG SPARROW *Melospiza melodia*

NORTHERN WATERTHRUSH

SWAMP SPARROW *Melospiza georgi*

MARSH WREN *Cistothorus palustris*

CAROLINA WREN *Thryothorus ludovicianus*

WHITE-THROATED SPARROW

COMMON GRACKLE *Quiscalus quiscula*

AMERICAN GOLDFINCH *Spinus tristis*

TREE SPARROW *Spizelloides arborea*

OAT-TAILED GRACKLE *Quiscalus major*

PROTHONOTARY WARBLER *P. citrea*

TUFTED TITMOUSE *Baeolophus bicolor*

ACK-CAPPED CHICKADEE

MOURNING DOVE *Zenaida macroura*

AMERICAN ROBIN *Turdus migratorius*

EASTERN COTTONTAIL *Sylvilagus floridanus*

AMERICAN MINK *Neovison vison*

NORTH AMERICAN PORCUPINE

Muskrats use their tails to propel themselves and to steer.

MUSKRAT *Ondatra zibethicus*

Beavers use their tails to steer and their webbed hind paws to propel themselves.

AMERICAN BEAVER *Castor canadensis*

RICAN BLACK BEAR *Ursus americanus*

EASTERN GRAY SQUIRREL

MUSKRAT *Ondatra zibethicus*

Otters use their tails and paws to propel themselves and to steer.

NORTH AMERICAN RIVER OTTER *Lontra canadensis*

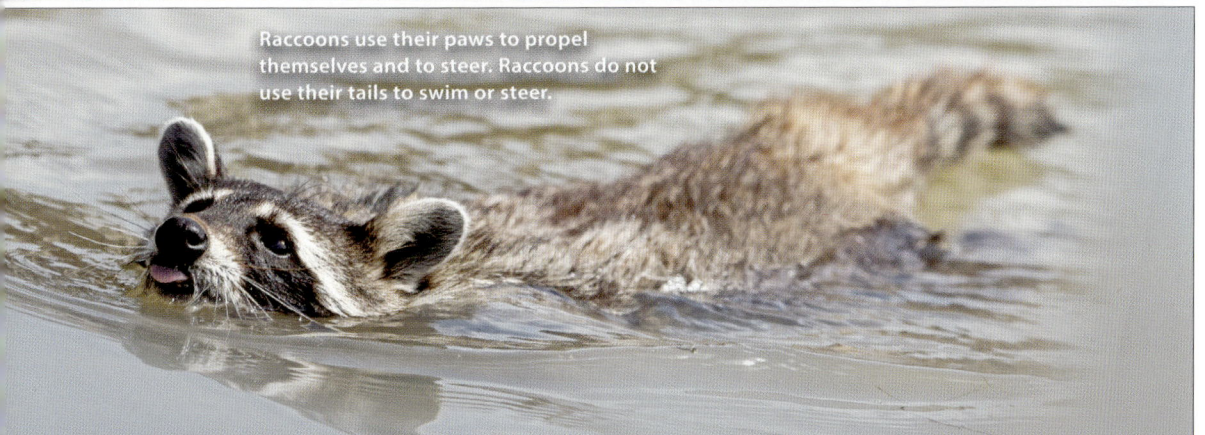

Raccoons use their paws to propel themselves and to steer. Raccoons do not use their tails to swim or steer.

RACCOON *Procyon lotor*

Next to a pond, a mixed grassy and emergent marsh displays a rich combination of Tussock Sedge, Northern Wild Rice, Tall Cordgrass, Royal Fern, Cinnamon Fern, and Common Cattail (not yet in bloom). Duckweed spots the water surface, and skinny alder shrubs mix in on the slightly higher ground.

Freshwater Marshes

Natural or older human-created ponds almost always have other types of wetlands associated with them. We draw lines around certain environments to better understand their ecology and biology, but nature abhors hard borders. In the real world, wetlands can defy easy classification. Is that pond edge a wet meadow, a marsh, or at least partly a shrub swamp? Often, it is all three at once, at least in small areas. However, many wetland environments are consistent enough in their character that classification can be a useful tool for understanding the unique plants and animals of the wetland types that surround ponds. This chapter and those that follow look at the most common types of wetlands and a few special kinds of ponds.

What is a marsh?

A marsh is a wetland dominated by emergent vegetation, usually grasses, sedges, rushes, and herbaceous species whose roots can withstand constant immersion in water. Ecologists often distinguish between marshes with deep water (averaging four to 15 inches of standing water) versus shallow-water marshes or wet meadows (*see illustration, p. 6*).

The wide variety of plants and animals living in or near marshes makes freshwater marshes some of the most diverse and beneficial natural environments. Marshes are incredibly productive ecosystems because the dominant grassy and herbaceous plant leaves wither and die back each winter. A large proportion of the plant nutrients are recycled back into the marsh environment each year. Most marsh grasses, sedges, and rushes are perennial plants that overwinter and spread through rhizomes or underground stems. The plants lose their leaves but then sprout again from rhizomes in the spring.

Marsh waters are typically well oxygenated because shallow waters have a high surface-to-volume ratio for gas exchange, preventing stag-

The Pied-Billed Grebe (*Podilymbus podiceps*), a small waterbird, thrives in the freshwater wetlands of the Eastern United States. The Pied-Billed Grebe favors emergent marshes near quiet ponds, where abundant cattails, grasses, and wild rice provide both material and shelter for their floating nests. This grebe's diet is primarily small fish, insects, and crustaceans, which are plentiful in emergent marshes. The sheltered, marshy environments of wetlands and small ponds are ideal for the grebe's elusive behavior, allowing it to dive and disappear from sight at the slightest hint of danger, a distinctive survival tactic.

nation. Dense marsh plant leaves, stems, and surface root systems act as a giant sponge and filter for waterborne silt and nutrients. The typically clear and well-oxygenated marsh water supports a wide range of animal life. However, the water-logged soils and peat under the marsh are poor in oxygen content, and this low-oxygen soil environment prevents most forest and field plants from colonizing marshlands. Only plants with adaptations to bring oxygen to their root systems can survive in wet, marshy soils.

Zonation in marshes

Freshwater marshes are rarely a uniform mix of one plant community from edge to edge. Most marshes show three distinct but sometimes intermixed zones of plants based on average water depth. Emergent plants such as cattails, Northern Wild Rice, Common Reeds, and other taller grasses dominate in the deepest marsh waters. The second deeper zone is the wet meadow or shallow marsh, dominated by grass-like plants (true grasses, rushes, and sedges). The third shrub-dominated marsh zone has the shallowest water; through succession, this shrub zone eventually evolves into a shrub swamp. Wet-tolerant shrubs such as Smooth Alder, Buttonbush, Redosier Dogwood, Black Willow, and Highbush Blueberry ring the shallow marsh edges.

Shallow marshes and wet meadows

Shallow marshes are dominated by grasslike plants (grass, sedges, rushes) mixed with many herbaceous species. The maximum water

An emergent marsh dominated by Northern Wild Rice (foreground center), with mixed Common and Narrow-Leaved Cattails and more Northern Wild Rice in the distance. Along the raised roadside at the bottom of the image are Bur-Marigold (Beggar's Tick, yellow flowers) and Purple Loosestrife spikes with flowers that have gone to seed.

depth may only be a few inches to a foot, but that is often sufficient to limit the growth of trees and shrubs. Plants that dominate low marshes and wet meadows must tolerate a wide range of water depths, but the flooding fluctuations increase the marsh's nutrient richness. In low-water periods, old leaves and dead plant materials break down more quickly in the presence of oxygen from the air, and these nutrients are recycled promptly into the marsh environment. Streams usually feed marshes, but most marshes also receive groundwater and rainwater runoff from the uplands around the marsh, and the runoff water is often rich in mineral nutrients.

Wet meadows (also called sedge meadows) are freshwater marshes with very shallow standing water for some of the year. Still, they typically lose much of their standing water in the heat and dryness of late summer and early autumn. Many wet meadows are dominated by dense clumps of Tussock Sedges, which can function almost like little islands in the marsh, particularly in the wetter seasons. Red-Winged Blackbirds often use the center of large Tussock Sedges as dry platforms to build their nests well above the marsh water level.

The edges of wet meadows can be excellent places for watching birds, butterflies, and dragonflies. Wet meadow wildflowers such as Joe-Pye Weed, Ironweed, Blue Vervain, and Swamp Milkweed are particularly attractive to larger, easily spotted butterflies including Monarchs, swallowtails, fritillaries, and buckeyes.

Spotted Joe-Pye Weed
Eutrochium maculatum

Tussock Sedge
Carex stricta

Eastern Tiger Swallowtail
Papilio glaucus

Wet Meadows—An Ephemeral Spring Habitat

Skunk Cabbage

Bluets

Swamp Buttercup

Tussock Sedge

Green Frog

Wet meadows in the Northeastern United States are ecologically rich ecosystems characterized by their saturated soil conditions, which are typically found in shallow basins, flat plains that become waterlogged in the springtime, or along the edges of streams and ponds. Wet meadows support a high diversity of plant life, including various sedges, rushes, and grasses, as well as colorful wildflowers such as Ironweed and Swamp Milkweed. The wet soil conditions, combined with abundant plant life, create a prime habitat for a variety of wildlife, including amphibians such as frogs and salamanders, birds such as Marsh Wrens and Red-Winged Blackbirds, and numerous insects.

Joe-Pye Weed is a common wildflower in Northeastern wetland edges and grassy fields. The light pink flowers attract butterflies, such as the Eastern Tiger Swallowtail (*see illustrations, previous page*). Dragonflies also seem to like hanging around Joe-Pye Weed, probably to pick off smaller pollinating insects.

With their abundant insect life, shallow marshes and wet meadows are a paradise for our common frog species: Bullfrogs, Green Frogs, Northern Leopard Frogs, Pickerel Frogs, Gray Treefrogs, and Spring Peepers. Marshes attract a wide range of bird species. Long-legged waterbirds including herons, rails, and the large but secretive American Bittern are all characteristic species of marshes and wet meadows. Smaller birds such as Song Sparrows, Swamp Sparrows, and Marsh Wrens are common in freshwater marshes.

The abundant wildlife, wide open spaces of wet meadows, and shallow marshes attract predators. Nocturnal raptors such as Eastern Screech Owls and Barred Owls will happily take frogs, toads, Meadow Voles (wild mice), and other small animals. Northern Harriers are frequently seen over wet meadows in spring and fall migration seasons. But the most commonly seen predator over grassy marshes might be the Red-Tailed Hawk. It's rare to spend time in wetland meadows and not see a Red-Tail flying in broad circles overhead.

Emergent marshes

Taller grasses such as Broad-Leaved Cattails, Northern Wild Rice, and Common Reed and larger sedges including Tussock Sedge, American Bur-Reed, and Giant Bur-Reed usually dominate deeper marshes. Mixed among these taller grasslike plants are wetland shrubs such as alders and Buttonwood. The Royal Fern is a common inhabitant of both shallow and deeper marshes. Pickerelweed and Arrow Arum are common in shallow emergent wetlands, and both plants seem to favor the edges of creeks and open water spots within emergent marshes.

Marshes often have scattered small Red Maples, but the maples are typically dwarfed due to their constant immersion in standing water and seldom reach the mature sizes seen in drier maple swamps. Even marshes dominated by grasses and other grasslike plants will have scattered wetland shrubs, including Smooth and Speckled Alder, Silky Dogwood, and Buttonbush. The scattered woody shrub species may indicate the first stages of succession from a grass-dominated marsh to a shrub swamp. As woody wetland shrubs become more numerous, the shrubs shade and outcompete their lower grasslike neighbors, and

Emergent freshwater marshes can be vast, as in this Lord Cove area along the Connecticut River in Lyme, Connecticut. Dominated by hundreds of acres of Narrow-Leaved Cattails, Lord Cove is also a rich complex of ponds and marsh streams, bordered by smaller shrub swamps and hardwood swamp wetlands along the creeks and small rivers that enter the cove.

Freshwater Marsh Habitat
Little Pond, White Memorial Foundation, Litchfield, CT

Great
Blue Heron

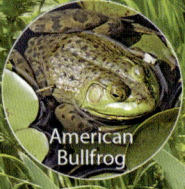

American
Bullfrog

Wood Duck

Mallard

Eastern
Screech Owl

Pickerelweed

White
Water-Lily

Broadleaf
Cattail

Great
Bulrush

Arrow Arum

Snapping
Turtle

Red-Winged Blackbirds, male and female

Tussock Sedge

Yellow Water-Lily

Northern Blue Flag Iris

Blue Dasher

Common Whitetail

American Kestrel

Arrow Arum

Common Muskrat

Yellow Flag Iris
Invasive

Virginia Rail

Purple Loosestrife
Invasive

Arrow Arum

American Bittern
Botaurus lentiginosus

Sora
Porzana carolina

Virginia Rail
Rallus limicola

eventually, what was marsh will evolve into a shrub swamp. In more extensive marshes, you'll often see a mix of communities in the same area: a grass-dominated marsh or wet meadow with patches of shrub swamp growing on the drier ground around the edges of the marsh.

Marshes dominated by grasses, sedges, and rushes are home to marsh specialists such as the American Bittern, Virginia Rail, and Sora (a small rail). Many heron species also feed and sometimes breed in freshwater marshes. And, of course, wetlands are full of dragonflies, damselflies, and other flying and aquatic insects. Joe-Pye Weed is a widespread wildflower in marshes and wet meadows, where its light-violet flowers attract many species of butterflies, dragonflies, and other insects.

Red-Winged Blackbirds inhabit open areas such as marshes, meadows, fields, and wetlands. In summer, Red-Wings feed mainly on insects and eat grains, berries, and other small fruits. In winter, their primary diet shifts to seeds, stray corn kernels in farm fields, and berries and nuts. Breeding season begins in early spring and continues through early summer. During this time, males establish and defend their territories and attract females with their unique "cong-ga-reeee!" calls, one of the most familiar sounds in all wetlands. Females typically build cup-shaped nests in low shrubs at the periphery of marshes or trees and lay two to five eggs per clutch. Within wetlands, Red-Wings also build nests in the centers of large Tussock Sedges. The young birds leave the nest after about two weeks and can fly after another two weeks.

Deeper marshes almost always have patches of open water, often near the center of the marsh. These open patches of water make ideal feeding grounds for dabbling ducks such as Mallards, American Wigeons, Blue-Winged Teals, and Northern Pintails. Dabbling duck species do not dive underwater to feed, and the shallow water of marshes offers them rich feeding at just the right depth. Open marshes also offer ducks and other birds protection from predators such as foxes and coyotes, which have difficulty moving in deeper marshes. Marshes, open wetlands, and farm fields along Eastern U.S. waterways offer essential feeding and resting areas in the spring and fall migration seasons.

Cattail marshes

Almost all deeper freshwater marshes have some cattail species present. Stands of Cattail often form a scattered mosaic alongside patches of Tussock Sedge or other marsh grasses. But some emergent marshes are so dominated by cattail species that they form a distinct marsh ecosystem. The most common cattail in freshwater marshes is the Broad-Leaved or Common Cattail, which forms dense stands that typically exclude other native plants except for scattered Northern

Wild Rice or Big Cordgrass. Unfortunately, the stands of cattails are often mixed with the invasive Common Reed (*Phragmites*). Narrow-Leaved Cattails are more tolerant of brackish water and are often found along the edges of brackish and salt marshes on the East Coast.

Narrow-Leaved Cattails are also much more tolerant of water polluted by sewage. Thus, a stand of Narrow-Leaved Cattails in inland areas may indicate a problem with runoff waters polluted by sewage or excess nutrients from lawn fertilizers and household soaps. Narrow-Leaved Cattail stands in roadside ditches may also indicate salt pollution from wintertime salt washing off roads. The Narrow-Leaved Cattail and the invasive Common Reed (*Phragmites*) tolerate the alkaline waters created by excess salts and pollutants, so you'll often see the two types of grass in mixed stands along roadsides. However, some Narrow-Leaved Cattail stands are perfectly natural and are indicators of neutral to basic soil types.

The Red-Winged Blackbird is found throughout North and Central America and is very common in wetlands in the Connecticut River region. Red-Wings are named for the male's brilliant scarlet epaulets against a deep black body. The brown, mottled, Red-Wing females look like stocky, oversized sparrows.

Red-Winged Blackbirds
Agelaius phoeniceus

The Emergent Zone of Ponds and Marshes
The life in and around cattails and grassy ponds

GREEN
DARNER
3 in.

Yellow
Water-Lily

MUSKRAT
16–28 in.

AMERICAN
BULLFROG
7–8 in.

PAINTED
TURTLE
5–10 in.

NORTHERN
PIKE
16–18 in.

BLUEGILL
4–8 in.

GREEN HERON
25–27 in.

Animals are not shown to scale

PUMPKINSEED
3–6 in.

AMERICAN
BULLFROG
TADPOLE
3–6 in.

REDBREAST
SUNFISH
4–6 in.

YELLOW PERCH
7–9 in.

Beaver workings are pretty easy to spot. Favorite wetland or riparian trees include aspens, cottonwoods, birches, and willows.

Beavers and wetlands

If we could travel back to the precontact East Coast of the 1500s, we might barely recognize many of the familiar rivers and streams in our landscape today. Streams and small rivers back then were more like chains of shallow lakes impounded behind tangled wooden dams. Extensive freshwater marshes and wet meadows broke up the forest cover, and few streams flowed for more than a mile or two before entering a pond or deep marshland. The profusion of freshwater wetlands supported vast numbers of waterfowl and other wetland inhabitants, and countless migratory fish brought the nutrients of the sea to the farthest inland brooks and streams. This rich but very wet landscape was heavily engineered, but not by humans.

Biologists call Beavers a keystone species, meaning they profoundly impact their ecosystem by creating habitats for many other species. Beavers modify streams and wetlands by building dams, creating new ponds, marshes, and swamps, increasing natural water storage and improving water quality.

Before 1620 and the arrival of European settlers, there were an estimated 400 million Beavers in North America. By 1900, the Beaver population in the Lower 48 States had dwindled to around 100,000, thanks to relentless hunting and trapping of Beavers for their fur. North America has lost half its wetland acreage since 1620, but most of those vanished marshes and swamps were not filled or bulldozed by

Wetland mammals are often mistaken for one another when swimming. There are very substantial size and weight differences among the common mammals, so it pays to be skeptical when you hear stories of "giant swimming rats" in the local park. Muskrats, beavers, and even otters are more widespread than you think, even in urban park wetlands.

NORTHERN RACCOON
Procyon lotor
28 in. long, 10–20 lbs.

Muskrats have horizontally flattened tails, which they use to swim.

MUSKRAT
Ondatra zibethicus
20 in. long, 3–4 lbs.

BROWN RAT
Rattus norvegicus
15 in. long, .75 lbs.

people. When trappers and fur hunters extirpated the Beaver, we also lost the Beaver's works. Within a few decades, the ponds and marshes behind Beaver dams disappeared as the barriers disintegrated and washed away.

By 1900, North America's largest rodent was absent from New England and nearly wiped out from the Eastern United States. In the early 20th century, enlightened naturalists and foresters realized how much wetland habitat diversity had been lost with the disappearance of Beavers. From the 1910s to the 1930s, small numbers of Beavers were reintroduced to Connecticut, Massachusetts, Vermont, and New Hampshire, where they had been absent for decades. The reintroduced Beaver population increased rapidly. Abandoned farmland was returning to the forest, and the new Beavers had no competition. Today there are no hard estimates for the Beaver population of the Eastern United States. Still, wildlife biologists estimate that the Beaver population has almost reached the carrying capacity of current wetlands. This is a far lower population than before 1620, but our much-reduced wetland acres now hold about as many Beavers as our

Two recent books on beavers and their importance to both human and natural history:

Eager: The Surprising, Secret Life of Beavers and Why They Matter, by Ben Goldfarb (White River Junction, VT: Chelsea Green, 2018).

Beaverland: How One Weird Rodent Made America, by Leila Philip (New York: Twelve, Hachette, 2022).

Beavers have vertically flattened tails and use their tails only to steer, not to paddle.

AMERICAN BEAVER
Castor canadensis
40–42 in. long, 40–70 lbs.

NORTHERN RIVER OTTER
Lontra canadensis
40–42 in. long, 10–33 lbs.

We can thank the North American Beaver (*Castor canadensis*) for this beautiful cattail marsh along Webster Road in Litchfield, Connecticut. The beavers have long dammed up the Miry Brook, but because the brook is small, the low water flows of summer result more in marshes than in open water. The emergent marsh is mostly Common Cattail (*Typha latifolia*), with Tussock Sedges (*Carex stricta*) to the right and a Red Maple (*Acer rubrum*), a classic wetland tree species, at the left.

Mature beaver dams are massive structures that the animals must constantly maintain to prevent leaks. Beavers are said to be extremely sensitive to the sounds of flowing water—a warning that their dam may be leaking.

current streams, ponds, and riverine forests can support. Beavers can't repopulate the wetlands that have vanished since 1620, but steadily, today's Beavers are rebuilding what was lost.

Beavers dam streams to create ponds deep enough to not freeze entirely in the winter. Swimming under the ice gives Beavers protection from predators, shelter from the elements, and a safe place to raise their young. Beavers build their lodges in water deep enough for the animals to enter and leave the lodge while underwater.

Beavers are sophisticated water engineers. In building their dams, Beavers generally choose locations where a natural narrowing of a stream occurs, such as between rock formations or where the stream enters a narrowing valley. This efficient siting allows the dam to be as short as possible and as high as needed to dam ponds to at least four to five feet deep.

The reintroduction of Beavers has had many ecological and direct benefits to human health and safety. Beaver ponds create more numerous and diverse wetlands than we might have otherwise, and these wetlands act as both a filter for pollutants and a sponge to regulate water flows. In droughts, the Beaver ponds provide a steady supply of water. In storms, the Beaver ponds and dams help buffer the flow of destructive stormwater. Ecologists call these benefits ecosystem services and assign dollar values to habitat restoration, pollution control, water regulation, and stormwater management. Beavers provide many millions of dollars in ecosystem services, and the damage from the Beaver's nuisance flooding is a tiny fraction of the benefits we receive.

Beaver lodges are large structures six to eight feet wide and have many large branches on the exterior. **Muskrat lodges,** though similar, are smaller and are made with the leaves of aquatic plants, with only a few smaller sticks showing.

Beaver lodge

Muskrat lodge

Beavers don't just create ponds. This small, four-acre wetland exists behind a Beaver dam and is a complex amalgam of pond, small marshes, cattail areas, and hardwood swamp in the background. This rich and varied environment is an ideal home for all kinds of forest and wetland plants and animals, all dependent on a keystone species: the North American Beaver.

A vibrant, healthy freshwater marsh dominated by typical marsh plants including Tussock Sedges, Pickerelweed, Arrowhead, Yellow Flag and Harlequin Blueflag irises, small Buttonbush and alder shrubs, and many small Red Maples. The maples are sapling-sized but are probably older trees that are dwarfed by the very wet soils. The small areas of open water increase the structural complexity of the marsh by giving animals such as frogs and waterbirds an edge from which to observe and hunt for prey. The open water areas are also highly attractive to flocks of waterfowl in spring and fall migration.

Little Pond marsh and boardwalk,
White Memorial Conservation Center, Litchfield, Connecticut.

Common plants and animals of northern freshwater marshes.

COMMON CATTAIL *Typha latifolia*

NARROW-LEAVED CATTAIL *Typha angustifolia*

An invasive species

COMMON REED *Phragmites austral*

SOFTSTEM BULRUSH *S. tabernaemontani*

PICKERELWEED *Pontederia cordata*

WHITE WATER-LILY *Nymphaea albe*

YELLOW WATER-LILY *Nuphar lutea*

COMMON DUCKWEED *Lemna minor*

ARROWHEAD *Sagittaria latifolia*

See also the ID plates in "Common Freshwater Plants and Animals."

REEN FROG *Lithobates clamitans*

SNAPPING TURTLE *Chelydra serpentina*

PAINTED TURTLE *Chrysemys picta*

ANADA GOOSE *Branta canadensis*

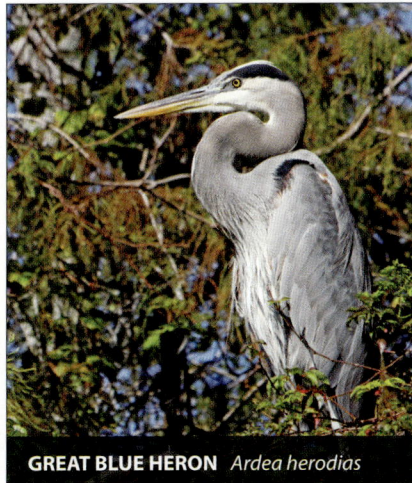

GREAT BLUE HERON *Ardea herodias*

RED-WINGED BLACKBIRD *Agelaius phoenicus*

ARSH WREN *Cistothorus palustris*

MUSKRAT *Ondatra zibethicus*

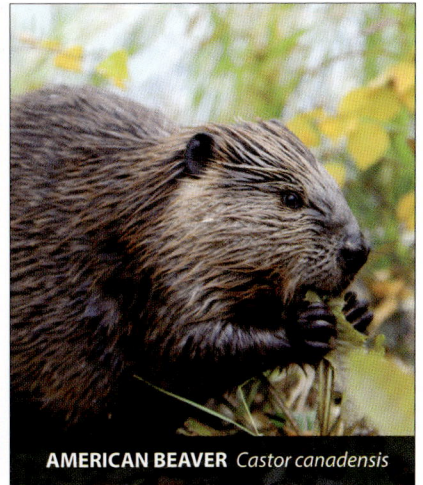

AMERICAN BEAVER *Castor canadensis*

Common plants and animals of southern freshwater marshes.

SAWGRASS *Cladium jamaicense*

COMMON CATTAIL *Typha latifolia*

PICKERELWEED *Pontederia cordata*

ALLIGATOR FLAG *Thalia geniculata*

MAIDENCANE *Panicum hemitomon*

BUTTONBUSH *Cephalanthus occiden*

SOUTHERN SWAMP LILY *Crinum americanum*

ARROWHEAD *Sagittaria latifolia*

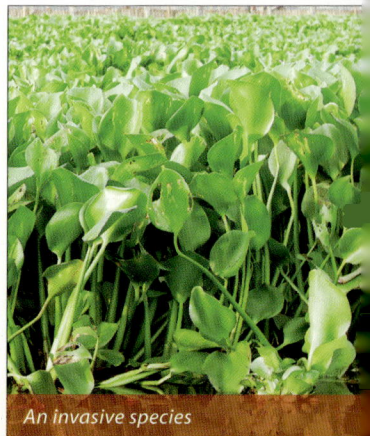

An invasive species

WATER HYACINTH *Eichhornia crassi*

See also the ID plates in "Common Freshwater Plants and Animals."

ORIDA GAR *Lepisosteus platyrhincus*

MOSQUITOFISH *Gambusia affinis*

AMERICAN BULLFROG *Lithobates catesbeiana*

REEN TREEFROG *Hyla cinerea*

SNAPPING TURTLE *Chelydra serpentina*

ALLIGATOR *Alligator mississippiensis*

NHINGA *Anhinga anhinga*

GREAT EGRET *Ardea alba*

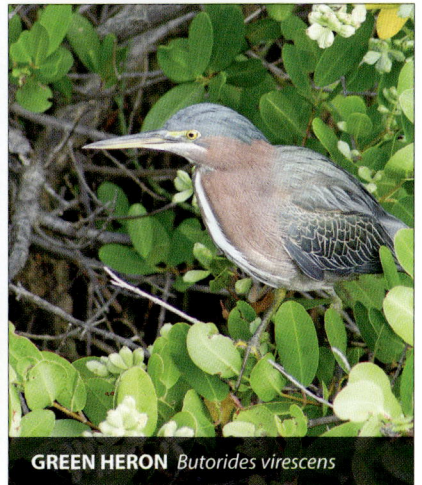
GREEN HERON *Butorides virescens*

Shrub swamps in the Northeastern U.S. are usually dominated by Smooth Alder, Sweet Pepperbush, Highbush Blueberry, Inkberry, Japanese Honeysuckle, and Black Willow shrubs, with an understory of wetland herbaceous plants such as Skunk Cabbage and ferns including Cinnamon Fern. In the foreground here are Tussock Sedges. Young Red Maples provide a shady overstory, and eventually the shrub swamp becomes a maple swamp of tall trees on moist ground.

Freshwater Swamps and Shrub Swamps

A swamp is a wetland dominated by large trees or substantial shrubs. Swamps are often found next to woodland streams or on older river floodplains, near ponds and lake edges where the shoreline rises gradually. They are found where there is level ground and the local water table is often at or a few inches above ground level. Hardwood swamps are often a late successional stage of earlier shrub swamps, and in most hardwood swamps, woody shrubs form a dense understory layer above the wet soil. Swamps are often seasonally flooded but are sometimes drier-looking, especially in late summer. A mature hardwood swamp may be challenging to distinguish from a floodplain forest when the ground is dry in August and September. Even then, however, true swamps often have a dense carpet of mosses and wetland ferns including Cinnamon Fern, Sensitive Fern, and Lady Fern. In contrast, drier floodplain forests show only occasional patches of moss and upland ferns such as Bracken Fern and Ostrich Fern.

You can recognize a mature swamp by the combination of shallow standing water under large wet-tolerant tree species including Red Maple, Black Tupelo, Yellow Birch, Red Spruce, Black Spruce, and Larch. Most hardwood swamps have a well-developed shrub layer of wetland species such as Sweet Pepperbush, Highbush Blueberry, Sheep Laurel, Leatherleaf, and Northern Arrowwood. A dead giveaway for swamp habitat is dense patches of Skunk Cabbage under mature trees, as Skunk Cabbages favor areas where the soil is saturated for much of the year. The drier edges of swamps often have a mix of Eastern Hemlocks, Red Maples, and various birch species, often with a dense understory of Mountain Laurel.

Swamps are critical wildlife habitats because they are often difficult for people to move in or through. The dense, tangled understory shrubs and mucky soils make swamps a natural refuge for many bird and mammal species. Swamps next to rivers also create natural wildlife

Common Yellowthroats (*Geothlypis trichas*) have a strong affinity for wet and densely vegetated habitats, making pond edges, marshes, and swamps their preferred environments. The dense vegetation is crucial for the Yellowthroat's breeding and feeding, offering protection from predators and a rich supply of insects.

The dense tangle of shrub vegetation at the Red Maple Swamp in Connecticut's Chatfield Hollow State Park disguises a shallow pond below. Under the alders, willows, Buttonbush shrubs, and young Red Maple trees is an old pond that averages two to three feet in depth. Red Maples are common swamp trees because their roots can tolerate water-saturated soil. At Chatfield Hollow, an excellent boardwalk with interpretive signage leads you through an environment that would otherwise be unapproachable for most visitors.

Smooth Alder
Alnus serrulata

Buttonbush
Cephalanthus occidentalis

Redosier Dogwood
Cornus sericea

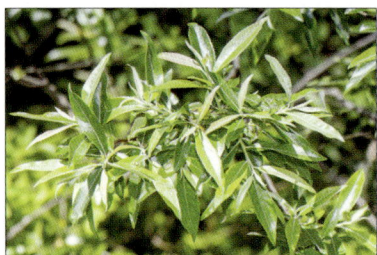

Black Willow
Salix nigra

corridors where wild animals can range over large areas while also having shelter from predators. Swamps make ideal nesting habitats for many warblers and other small songbirds, including the Northern Waterthrush, which nests only within hardwood swamps. Wood Ducks are one of the few dabbling duck species to nest in tree cavities, and hardwood swamps and adjacent ponds make perfect shelters and feeding habitats for both Wood Duck parents and their ducklings. Wooded swamps are also home to vernal pools (discussed below), a critical wetland wildlife habitat dependent on wet soil sheltered under a canopy of deciduous trees that shed their leaves in fall.

Shrub swamps

In the Northeast, a shrub swamp is a wetland dominated by woody plants such as Smooth Alder, Speckled Alder, Buttonbush, Silky Dogwood, Redosier Dogwood, and Black Willow. The alder species are particularly tolerant of standing water, and alders have tough, flexible branches that can withstand the force of flowing water in flood periods. Generally, a wetland is considered a shrub swamp when at least 50 percent of the area is covered with wetland shrub species. Water depth is variable with the seasons. The water table may drop below ground level in dry summers, although the soil remains saturated.

The Little Pond boardwalk at the White Memorial Conservation Center in Litchfield, Connecticut, provides outstanding views of a vast willow–alder–Buttonbush shrub swamp. The mile-long boardwalk passes through several kinds of emergent and shrub swamps, tree swamp areas, and a shallow open-water pond associated with the Bantam River (*see illustration, pp. 162–63*).

Shrub swamps are typically found in low-lying areas with poor drainage and are flooded by groundwater or surface water. They are often considered a transitional succession stage between wet meadows and true swamps, which are dominated by larger trees such as Red Maple, Black Tupelo, Yellow Birch, Black Ash, Balsam Fir, and Black Spruce. These dense, tangled environments provide important refuge habitats for many kinds of wildlife. Shrub swamps are also crucial in regional water filtration and help to reduce erosion and flooding in the surrounding area.

Shrub swamps evolve in wet bottomland areas that might once have been open ponds but are now filled with vegetation and shallow standing water. Some shrub swamps are seasonally dry, particularly from July through September in years with less rain than average. Small shrub swamp areas exist near most substantial ponds and streams, but Small's Swamp in the Pilgrim Heights area of Cape Cod National Seashore is a perfect place to explore this freshwater habitat (*see illustration, p. 168*). Small's Swamp sits in a classic kettle depression or kettle hole. Kettle holes formed the same way kettle ponds did (see "Kettle Ponds"), but the bottoms of kettle holes do not dip much below the local water table and so do not become open ponds. Common woody vines such as Catbrier, Virginia Creeper, and Poison Ivy are usually present, especially in the drier areas surrounding the wet swamp. This dense tangle of vegetation makes an ideal refuge, feeding, and breeding area for many small animals and birds. Even larger mammals such as Eastern Coyotes, Bobcats, White-Tailed Deer, and Black Bears find swamps inviting. Red Maples thrive in swampy areas and will invade shrub swamps, eventually becoming dominant and turning the shrub swamp into a Red Maple swamp, as at the excellent Red Maple Swamp Trail at the Fort Hill area of Cape Cod National Seashore or the Red Maple Swamp Boardwalk at Connecticut's Chatfield Hollow State Park (*see illustration, pp. 158–59*).

Atlantic White Cedar
Chamaecyparis thyoides

A rare form of swamp along the Atlantic Coast

The Atlantic White Cedar Swamp Trail at the Cape Cod National Seashore is one of the best and most accessible examples of a once-common but now rare wetland community—Atlantic White Cedar swamp. Such swamps were formerly common in coastal New England and on the Atlantic Coastal Plain south of New York Harbor. The dense, rot-resistant wood of the Atlantic White Cedar was highly prized as a house and shipbuilding material in the eighteenth and nineteenth centuries, so most cedar swamps were clear-cut long ago.

The boardwalk through the Atlantic White Cedar Swamp in the Marconi Station area of the Cape Cod National Seashore. This short hike (just over a mile) is probably the most accessible cedar swamp along the East Coast. It's best to visit cedar swamps in the drier and cooler months—in summer the swamps are full of mosquitoes, which breed abundantly in the still pools of the swamp floor.

Maple–Alder–Buttonbush-Dogwood Shrub Swamp
Bantam River marshes along Whites Wood Road, Litchfield, CT

Red-Tailed Hawk

Red Maple

Red Maple

Smooth Alder

Black Gum (Tupelo)

Buttonbush

Silky Dogwood

Mixed Tussock Sedges and Bulrushes

Mixed:
Smooth Alder
Buttonbush
Silky Dogwood
Royal Fern

Royal Fern

Buttonbush

Wild Rice

Tussock Sedge

CAROLINA
WREN

MARSH
WREN

Red Maples

GRAY
CATBIRD

Black Willow

WOOD
DUCKS

A view of the understory of a shrub swamp and the quiet stream that feeds the swamp. The dense tangle of shrub stems and roots makes shrub swamps not only ideal wildlife refuges but also ideal natural protection against sudden floods in rainstorms. In flood the whole swamp become a giant sponge to absorb stormwater. The complex stems and roots of shrub swamps also act as a natural speed brake against fast-flowing stormwater.

The freshwater swamps and ponds of the Nature Conservancy's Nags Head Woods Preserve in coastal North Carolina contain a wide variety of inhabitants, including such fish as Smallmouth Bass, Bluegill, Pumpkinseed, and Black Crappie. A small but elusive colony of River Otters lives in the preserve's pond and swamp areas. This pond is covered with duckweed in late summer. The preserve's complex of wetlands and upland coastal forest exists entirely on the sand of a barrier island in the Outer Banks.

Atlantic White Cedar swamps occur in old glacial kettle pond areas that have silted up or in other natural hollows in the landscape that hold saturated soils all year. The cedars are adapted to the swamp's water-saturated, acidic soils and form a dense canopy that shades the understory, creating a cool and moist environment even during the drier summer months. The terrain is often uneven and spongy, featuring peat moss that accumulates over time due to the slow decomposition rates in the acidic waterlogged conditions. In the Cape Cod Atlantic White Cedar Swamp, a sturdy boardwalk leads visitors through the swamp. White Cedar swamps are home to a diverse array of other plants. The understory is populated with various ferns, including the Cinnamon Fern and Royal Fern, which thrive in the shadowy dampness. Shrubs such as Highbush Blueberry and Swamp Azalea flourish here, adding to the biodiversity and serving as food sources for wildlife.

Coastal New Jersey's Pine Barrens region offers a number of surviving Atlantic White Cedar swamps with boardwalks and trails that make it possible to appreciate this rare habitat. Allaire State Park in northern New Jersey preserves a White Cedar swamp. The Great Egg Harbor River Wildlife Management Area and the Batsto River White Cedar swamp along the Delaware River near Bridgeport, New Jersey, are well-known cedar swamps with trails. Pocomoke State Forest, located

Bald Cypress swamps such as this one in Georgia's Savannah River National Wildlife Refuge are iconic wetland ecosystems found along the Southeastern U.S. coast, especially in Florida, Louisiana, Georgia, and the Carolinas. These swamps are characterized by the presence of the Bald Cypress (*Taxodium distichum*), a tree that is notable for its buttressed base with distinctive flared trunks and its prominent cypress "knees" that protrude from the water near the trunks. The cypress knees are thought to provide oxygen to the cypress root systems, which are usually submerged in standing water or saturated soil. Bald Cypress swamps occur in shallow areas with slow-moving waters, such as shallow coastal plains and river floodplains.

in coastal Maryland near Pocomoke City, is home to the northernmost Bald Cypress swamp on the East Coast and also contains areas of Atlantic White Cedar swamp.

Coastal swamps

Shrub and tree swamps are not just inland phenomena. Many Eastern Seaboard barrier islands and bay coasts contain at least small areas of various kinds of swamps. Aside from the Cape Cod and New Jersey areas mentioned above, southern Virginia and North Carolina's Outer Banks areas contain outstanding swamp areas located on the mainland shore and on barrier islands well offshore. The Nature Conservancy's Nags Head Woods Preserve contains outstanding examples of coastal tree swamps, pocosin wetland habitat (see "Northern Bogs and Southern Pocosins"), as well as coastal freshwater swamps along the shores of Albemarle Sound. Georgia's Savannah National Wild Refuge contains large areas of virgin Bald Cypress swamp (*see illustration, previous page*), as well as many other large wetland areas along the Savannah River. The Savannah River is a particularly attractive destination for birders and naturalists in the fall migration months. In autumn, the swamp trees fill with flocks of songbirds migrating to winter territories in the Caribbean and South America. The major Eastern duck and goose species use the vast Savannah River complex for feeding and shelter, and huge flocks of ducks, swans, and Snow Geese winter on the river marshes.

The Small's Swamp Trail in Truro, within the bounds of the Cape Cod National Seashore, is an outstanding way to visit a vibrant shrub swamp habitat without getting your feet soaked. The trail wanders down through a Pitch Pine forest and into a small kettle hollow that has soaked soil and bits of standing water in all but the driest parts of late summer and early fall. Boardwalks take you through the wetter parts of the swamp.

The Black Bear and wetland areas

You might not think about large wildlife such as the American Black Bear (*Ursus americanus*) as a wetland inhabitant, but shrub and tree swamps as well as wetlands along river corridors are crucial habitats for many kinds of larger wildlife. When you are the size of a Black Bear, it's hard to move across today's developed landscapes without attracting human attention. Rivers and other wetlands offer larger wildlife travel corridors to move about quietly. Wetlands also offer omnivores such as the Black Bear a rich variety of foods, including plant materials, berries, nuts, and many smaller animals and fish. In times of drought, wetlands are important water sources for all wildlife. In the heat of summer, wild ponds and their associated shrub swamps offer animals a chance to cool off without being noticed by people.

True natural wetlands—as opposed to man-made or manicured park ponds—are almost always some combination of wetland types. Here a small pond is surrounded by an emergent marsh on the right and a small area of Red Maple swamp on the left. The marsh probably reflects the gradual filling-in of the pond through sediments carried by a feeder stream that enters at the center rear of the image. The typical fate of all small ponds is to fill over time and become marshes or swamps.

Common Northern swamp species.

RED MAPLE *Acer rubrum*

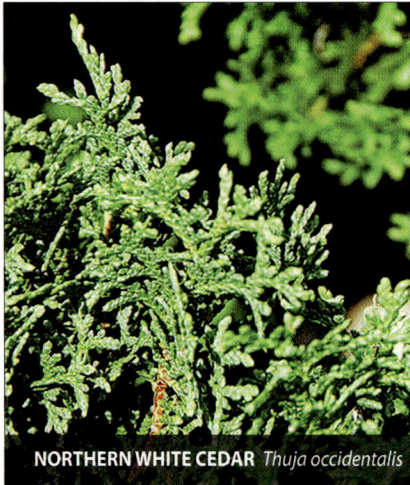
NORTHERN WHITE CEDAR *Thuja occidentalis*

BLACK SPRUCE *Picea mariana*

BALSAM FIR *Abies balsamea*

SWAMP AZALEA *Rhododendron viscosum*

HIGHBUSH BLUEBERRY *V. corymbo*

WINTERBERRY HOLLY *Ilex verticillata*

BUTTONBUSH *Cephalanthus occidentalis*

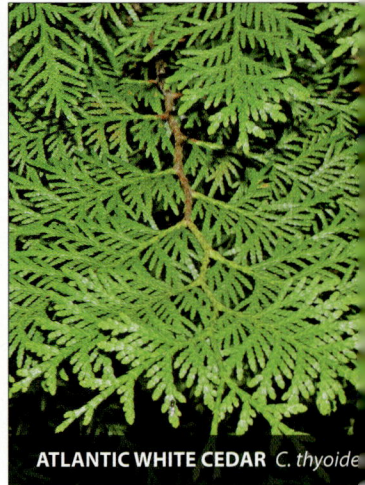
ATLANTIC WHITE CEDAR *C. thyoide*

See also the ID plates in "Common Freshwater Plants and Animals."

EEN FROG *Rana clamitans*

EASTERN NEWT *Notophthalmus viridescens*

SPOTTED SALAMANDER *A. maculatum*

AINTED TURTLE *Chrysemys picta*

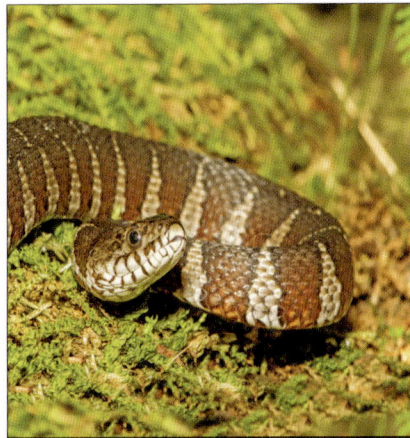

NORTHERN WATER SNAKE *Nerodia sipedon*

WOOD DUCK *Aix sponsa*

REAT BLUE HERON *Ardea herodias*

SWAMP SPARROW *Melospiza georgiana*

MUSKRAT *Ondatra zibethicus*

Common Southern swamp species.

BALD CYPRESS *Taxodium distichum*

WATER TUPELO *Nyssa aquatica*

RED MAPLE *Acer rubrum*

SWEETGUM *Liquidambar styraciflua*

SWAMP CHESTNUT OAK *Q. michauxii*

WILLOW OAK *Quercus phellos*

BUTTONBUSH *Cephalanthus occidentalis*

SWAMP DOGWOOD *Cornus foemina*

SWAMP AZALEA *Rhododendron visco*

See also the ID plates in "Common Freshwater Plants and Animals."

REEN TREE FROG *Hyla cinerea*

PIG FROG *Rana grylio*

TWO-TOED AMPHIUMA *A. means*

LLIGATOR *Alligator mississippiensis*

COTTONMOUTH *Agkistrodon piscivorus*

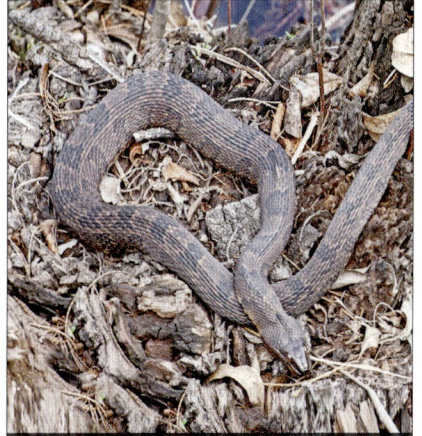

BROWN WATER SNAKE *N. taxispilota*

OOD DUCK *Aix sponsa*

PROTHONOTARY WARBLER *P. citrea*

SWAMP RABBIT *Sylvilagus aquaticus*

The leaves that line the bottoms of vernal pools are not just forest litter; they form the base of the food chain that makes life possible in these short-lived pools. Last year's leaves are a rich source of nutrients for the complex community of microorganisms inhabiting these pools. Every leaf is host to a biofilm (*see illustration, p. 33*) that breaks down the vegetal material. The microbial loop (*see illustration, p. 29*) then absorbs and recycles the leaf nutrients, making them available to the rich community of insects, amphipods, fairy shrimp, and insect larvae in the pool. The various small animals then provide the food for frogs, toads, and salamanders that breed there.

Vernal Pools: Seasonal Ponds

Vernal pools are small depressions or low spots in swamp forests, in floodplain forests, or near more extensive wetlands that fill with rainwater or snowmelt for at least several months in the spring. Vernal pools may be located near streams or ponds, but streams or groundwater flows do not feed the pools. The size and depth of vernal pools vary considerably, ranging from small puddles to larger pools spanning 10 feet or more. Deeper or more extensive vernal pools may persist for much of the year, but generally, vernal pools have no standing water in dry, hot summer weather. Often, vernal pool areas will refill in the wet fall months. Fall vernal pools, though less ecologically important or rich in life than springtime pools, are essential to some salamander species. Tiger Salamanders and Spotted Salamanders often mate and lay eggs in fall vernal pools, betting that the pools will reemerge in the spring and give their offspring a head start on development. As seasonal wetlands that disappear for at least part of every year, vernal pools are too short-lived to support fish, and this very lack of predatory, egg-eating fish is critical to the amphibians that depend on vernal pools for their mating and reproduction.

Most vernal pools depend on the shelter and shade of the forests around them, and forest pools will quickly dry up if the tree cover is lost due to development or logging. Once most of the trees are gone, the ground is exposed to heat and sun, and the vernal pools lose the soaked autumn leaves, biofilms, and abundant microorganisms that are the vital base of the pool food chain.

Unlike most wetland environments, vernal pools are defined by the animal communities that the pools support and not by characteristic vegetation. Most vernal pools are unvegetated within the pool, and their bottoms are typically lined with a layer of leaves that fell the previous autumn. These months-old rotting leaves host rich biofilms composed of bacteria and fungi. These biofilms form the base of

The Red Eft is the brightly colored juvenile stage of the common Eastern Newt (*Notophthalmus viridescens*), a species of salamander.

The newt's brightly colored skin is a warning—the skin produces tetrodotoxin. This potent neurotoxin is found in their skin and serves as a powerful deterrent against predation. Although Red Efts are not considered dangerous to humans, it's a sensible precaution to wash your hands after handling a Red Eft.

And, of course, don't eat one.

A classic vernal pool in early spring. This pool sits in a swampy bottomland forest area watered mostly by artesian leakage from the steep hillsides nearby. American Toads, Wood Frogs, Pickerel Frogs, and Spotted Salamanders all use this and nearby pools to mate and lay eggs. Vernal pools last until about midsummer, and by the normally dry season of early September, no standing water remains under the trees, although the soil stays moist.

Skunk Cabbage
Symplocarpus foetidus

Tussock Sedge
Carex stricta

American Toad
Anaxyrus americanus

Eastern Newt
Notophthalmus viridescens

the vernal pool food chain, and the films are consumed directly by fairy shrimp, amphipods, and insect larvae. Some vernal pools host a variety of typical wetland plants at or around their edges, including scattered Skunk Cabbages, Tussock Sedges, Highbush Blueberry, Red Maple saplings, and Buttonbush shrubs. However, most vernal pools are found in forests, which are in deep shade for much of the year, so plant life in and around vernal pools is not usually lush owing to the lack of light.

Wetland-adapted plants can help you spot vernal pool areas even when the pools have dried up in the summer heat. Skunk Cabbages and Tussock Sedges are particularly useful for spotting ephemeral wetlands even when the water has gone, since both plant species can survive long months of local drought.

Importance to forest amphibians

Vernal pools are a crucial habitat for breeding amphibians. Short-lived vernal pools can support frog and toad eggs and tadpoles that require about four months to mature. Most salamanders, however, require pools that persist for at least five months, so only larger vernal pools usually have salamander eggs and young. Although people mainly see vernal pools while on daytime hikes, dusk and night are when most of the critical activity happens in vernal pools. Protected by the dark from many predators, frogs, toads, and salamanders move across the forest floor into vernal pools to mate and lay eggs.

With species such as the Wood Frog, springtime migrations and mating can be noisy affairs even during the day and are sometimes deafening at night in peak season. To experience the full impact of spring in a vernal pool, try to locate a pool near enough to a road or park trail that you can safely visit it in the dark. Bring a good flashlight, wear sturdy waterproof boots, and avoid sudden movements or chatter with companions.

Ecological significance

Amphibians are hardly the only animal group to benefit from vernal pools. As in many other wetlands, the microbial loop (*see illustration, p. 29*) and biofilms (*see illustration, pp. 32–33*) play crucial roles in recycling the nutrients of forest leaves and plant litter and making this rich source of nutrients available to animals.

Tiny copepods, fairy shrimp, the larvae of many flies and dragonflies, and beetle larvae are prey animals for salamander larvae. They may appear in vast numbers, sometimes forming easily visible clouds of life in the water of vernal pools. Other invertebrates, such as beetle larvae, may turn the tables and prey on amphibian eggs or even on developing frog tadpoles. Some invertebrates, such as fairy shrimp (family name), have a unique life cycle adapted to the transient nature of vernal pools. The shrimp eggs can remain dormant in the bottom of the dry pool for months or even years, hatching only when the pool refills.

Amphibians are a crucial layer of the food chain in most forest and wetland ecosystems. Frogs, toads, and salamanders and their young feed on the invertebrates and insect larvae of vernal pools and the moist forest floor and in turn become food for raccoons, owls, foxes, night-herons, and other larger predators of wetlands.

In addition to their ecological significance, vernal pools provide valuable ecosystem services, including flood control, groundwater recharge, and nutrient cycling. Vernal pools and moist forest swamp areas act as natural sponges, absorbing excess water during heavy rains and releasing it slowly into the environment, reducing the risk of flooding.

Wood Frogs

The Wood Frog (*Lithobates sylvaticus, see illustration, pp. 186–87*) is an easily observable keystone species in vernal pools throughout the Northeastern and Atlantic Coast regions as far south as northern Virginia and the Blue Ridge Mountains. Although Wood Frogs spend most of the year as quiet, even stealthy inhabitants of moist forest floors, in early spring these frogs explode into an audible celebration of mating and egg-laying in vernal pools.

Wood Frogs are known for their remarkable frost tolerance. During winter, the frogs can survive with most of their body water frozen, thanks to cryoprotectants such as glucose in their vital organs and bloodstream. The formation of ice crystals in body tissue would be fatal to most animals. Wood Frogs, however, have evolved a combination of glucose and urea that acts as antifreeze in blood and tissue, protecting their cells from damage. Thus, Wood Frogs are usually the first amphibians to emerge and sing in the early spring, long before most other amphibians (*see illustration, p. 76*).

The Wood Frog is terrestrial for most of the year and can be found in various forested environments. However, these frogs prefer moist, deciduous forests with abundant leaf litter, where they feed on forest floor prey such as insects and other invertebrates. With a typical lifespan of three to five years, Wood Frogs are an essential prey item for avian and mammalian forest predators such as Black-Crowned Night-Herons, Raccoons, and Screech Owls.

The Wood Frog's adaptation to cold climates and ability to exploit temporary breeding sites mitigate competition and predation pressures and give the species a unique ecological niche and wide distribution throughout northern North America well beyond the range of most other amphibian species. This widespread distribution makes

Black-Crowned Night-Heron
Nycticorax nycticorax

Fairy Shrimp
Order Anostraca

Dragonfly larva
Order Odonata

Spring Peeper
Pseudacris crucifer

Spring and Autumn Vernal Pools
Temporary but critical wildlife communities

Mourning Cloak Butterfly, an early spring species

Eastern Comma Butterfly, an early spring species

American Toad

Spring Peeper

Gray Treefrog

Spotted Salamanders

Red-spotted Newt

Garter Snake

Wood Frog

Wood Frog Egg Mass

Wood Frog Tadpoles

Diving Beetle larva preying on a tadpole

Hatchling Eastern Painted Turtle

Common Snapping Turtle

Animals are not shown to scale

Common Darter Dragonfly

Skunk Cabbages

Skunk Cabbage Flower

Eastern Box Turtle

Green Frog

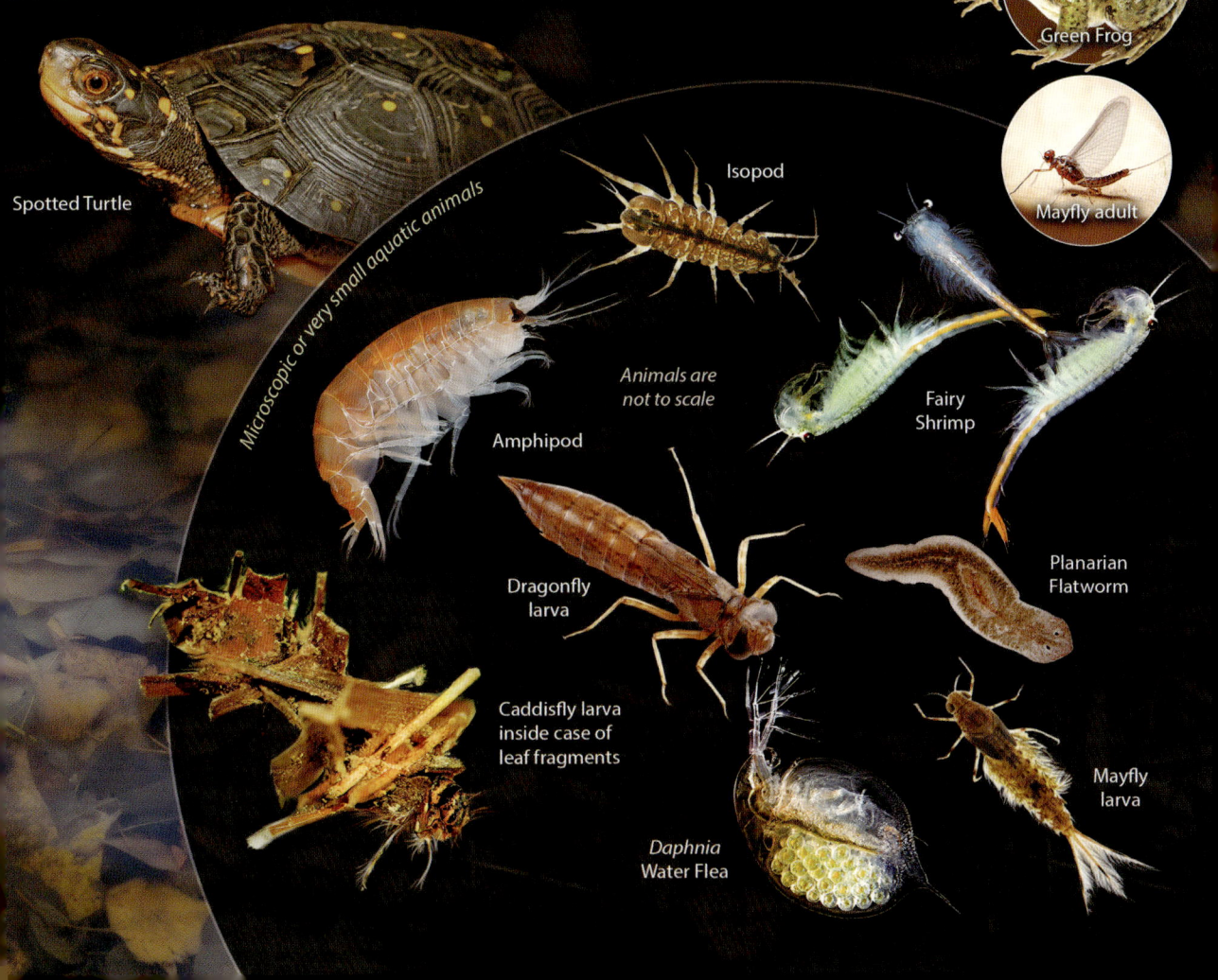

Spotted Turtle

Microscopic or very small aquatic animals

Isopod

Mayfly adult

Amphipod

Animals are not to scale

Fairy Shrimp

Dragonfly larva

Planarian Flatworm

Caddisfly larva inside case of leaf fragments

Daphnia Water Flea

Mayfly larva

Cinnamon Fern fiddleheads
Osmundastrum cinnamomeum

Sensitive Fern
Onoclea sensibilis

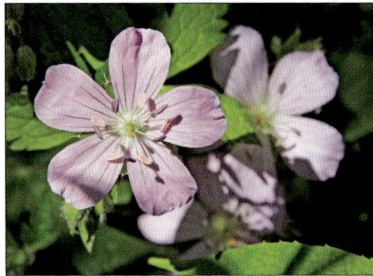

Wild Geranium
Geranium maculatum

the species an ideal subject for ecological studies on the impact of climate change, habitat loss, and the effects of habitat fragmentation.

Life history of Wood Frogs

Wood Frogs are among the first amphibians to emerge from hibernation in the spring. The onset of the frog's breeding season is triggered by rising temperatures and snow melting, which typically occurs from March to May, depending on latitude and altitude. Wood Frogs then migrate from their forest floor hibernation areas to vernal pools.

Upon the frogs' arrival at breeding sites, a frenetic and noisy mating activity commences. Males arrive first and begin calling to attract females. Their call is a distinctive, short clucking sound akin to a duck's quack. These vocalizations serve both to attract females and to establish dominance among males.

Wood Frog males are typically smaller than females and exhibit aggressive competition for mating opportunities. Wood Frogs engage in what is known as explosive breeding, where all mating activity occurs in a brief time-frame. This urgency is driven by the transient nature of the breeding sites, which may dry up quickly as the environment warms.

When a female chooses a male, he climbs onto her back in a mating grasp known as amplexus. The pair maintains this clasped position for up to several days until the female has released the eggs. The male fertilizes the large mass of eggs, usually between 1,000 and 3,000, as they are deposited in the water. The fertilized Wood Frog egg masses are typically attached to submerged vegetation and are jellylike, providing some protection from predators and temperature fluctuations.

Wood Frog eggs hatch into tadpoles within a few weeks, depending on the water temperature. Tadpoles are herbivorous, feeding on algae and plant residue in the water. Rapid development is crucial—the vernal pools can dry up within a few months after the last frost. Tadpoles typically metamorphose into froglets by mid- to late summer. These young frogs leave the water and spend the rest of the year in the forest, growing and maturing until they are ready to breed in the following year.

Wood Frogs' reproductive strategy is highly adapted to their cold, often unpredictable environments. The ability to breed quickly and in temporary pools reduces the risk of predation on their offspring and ensures that a new generation can mature even in short summers. Although a reliance on suitable weather conditions and the availability of temporary pools poses risks, this strategy has allowed Wood Frogs to thrive across a broad geographic range, showcasing their adaptability and resilience.

Even Atlantic Coast barrier islands contain freshwater wetlands and vernal pools. Here, at the Nags Head Woods Preserve on North Carolina's Outer Banks, the coastal forest contains many shallow ponds and vernal pools. The rich and varied amphibian life here includes American, Fowler's, and Spadefoot Toads, Spring Peepers, Bullfrogs, Southern Leopard Frogs, and Green Frogs (*see illustrations, pp. 114–15*). Marbled and Red-Backed Salamanders also breed in these vernal pools (*see illustrations, pp. 116–17*).

The First Frogs of Spring
Wood Frog (*Lithobates sylvaticus*) mating and reproduction

Wood Frogs are the earliest frogs to emerge from hibernation in spring. Triggered by the warming temperatures and melting snow, the adult frogs migrate to vernal pools, temporary bodies of water that form in the spring. The males arrive first, inflating their vocal sacs and creating a chorus of duck-like quacking calls to attract females. Once the females arrive, mating occurs, and the females lay large, spherical egg masses attached to submerged vegetation. After breeding, the adult frogs return to the surrounding woodlands, while the eggs hatch into tadpoles that develop within the vernal pool throughout the spring and summer.

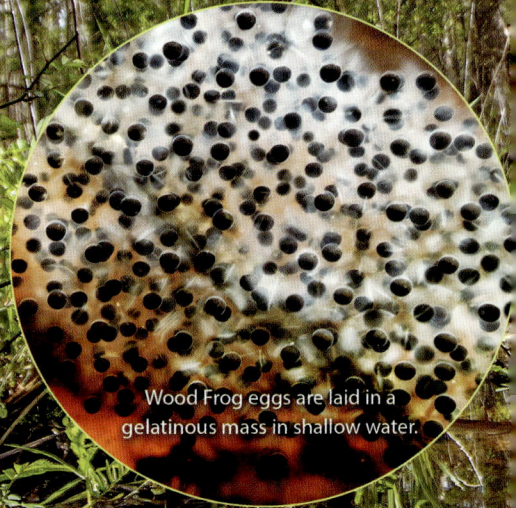

Wood Frog eggs are laid in a gelatinous mass in shallow water.

Springtime vernal pools are critical to breeding for many amphibian species because the pools are too short-lived to support fish. In larger permanent ponds the resident fish would eat most of the frog and salamander eggs and young.

An adult Wood Frog within a mass of maturing eggs. Frogs and other amphibians don't tend their eggs after laying, but you'll often see egg masses near mating frogs in vernal pools.

Wood Frog tadpoles

Adult Wood Frogs live most of their lives in moist woodlands, and only enter standing water in the springtime to mate and lay eggs.

Animals found in or near vernal pools in the Eastern United States.

PHANTOM MIDGE LARVA *Chaoborus sp.*

WATER STRIDER *family Gerridae*

WHIRLIGIG BEETLES *family Gyrinida*

BACKSWIMMER *family Notonectidae*

GIANT WATER BUG *Lethocerus americanus*

FAIRY SHRIMP *Eubranchipus sp.*

WATER BOATMAN *family Corixidae*

CADDISFLY LARVA *order Trichoptera*

DIVING BEETLE *family Dytiscidae*

AGONFLY NYMPH *order Odonata*

FINGERNAIL CLAMS *family Sphaeriidae*

EASTERN NEWT *Notophthalmus viridescens*

ASTERN NEWT "RED EFT"

JEFFERSON SALAMANDER *A. jeffersonianum*

GREEN FROG *Lithobates clamitans*

MERICAN TOAD *Anaxyrus americanus*

SPRING PEEPER *Pseudacris crucifer*

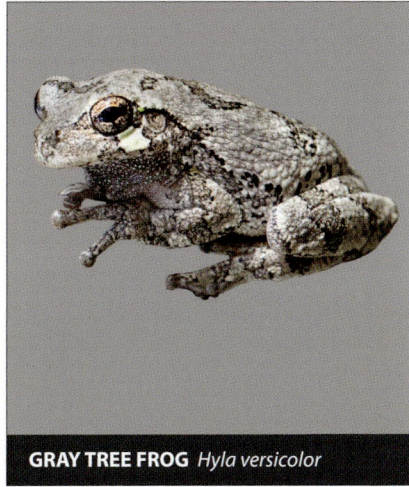

GRAY TREE FROG *Hyla versicolor*

A flower of the Purple Pitcher Plant (*Sarracenia purpurea*) found in cool acidic wetlands and bogs throughout North America. Its unique ecology and adaptation strategies allow it to thrive in nutrient-poor, acidic soils where other plants struggle. Bog areas typically lack the necessary elements for plant growth due to the leaching of essential nutrients such as nitrogen and phosphorus by acidic water. By trapping and digesting insects, pitcher plants can access a consistent and reliable nutrient source directly, enabling them to colonize areas that might otherwise be inhospitable.

Pitcher plants typically flower in early spring, often before the full development of its insect-catching pitcher leaves. The flowers are held high on a stalk above the pitchers to avoid trapping potential pollinator insects such as bees and flies. This vertical separation ensures that the plant can reproduce while still feeding effectively. Later in summer, after the flowers have faded, the sweet nectar in the pitcher leaves attracts and traps bees and flies.

NORTHERN BOGS AND SOUTHERN POCOSINS

Bogs are one of the rarest wetland environments in the Northeastern and Atlantic Coast regions. In northern New England, bog environments can form along the shallow edges of ponds and lakes, but they mostly form in cool, moist microclimates such as isolated valleys and glacial kettlehole ponds. These "frost pocket" or bottomland bog ponds are fed by rainwater and sometimes by small, low-volume streams, and they typically do not have an outlet stream. Plant decay in bogs is prolonged, and bog water is very acidic and lacking in nutrients. The dominant sphagnum moss cover builds up over centuries, converting a previously open pond into a layered basin filled with living moss on top, followed by some depth of older, dead sphagnum peat, and then some depth of open water beneath (*see illustration, overleaf*).

Most Northeastern bogs are remnants of the end of the Wisconsinan Glaciation (the Ice Age) 20,000–15,000 years ago, when the environment was more like that of northern Canada today. Arctic heaths and bog environments survived the end of the Ice Age in isolated, cold pocket valleys or on mountaintops. With their highly acidic water, specialized flora, and poor nutrient supply, bogs attract few animal species besides insects. Still, their unique flora makes bogs a fascinating environment to visit.

The Hawley Bog in west-central Massachusetts is a spectacular and easily accessible example of a bog environment, with good trails to the bog and a long boardwalk that takes you from the upland edges well out onto the sphagnum surface of the bog. All the major bog plant species are well represented, and the boardwalk also passes by Potash Brook, which feeds the Hawley Bog. Potash Brook is a small but fascinating transitional area with characteristics of a conventional freshwater wetland along with more typical bog shrubs and trees as the brook passes into the actual bog area (*see illustration, pp. 196–97*).

The Round-Leaved Sundew (*Drosera rotundifolia*) is a small carnivorous plant that inhabits the bogs and wetlands of the Northeastern United States. Sundews are particularly adapted to thrive in the nutrient-poor, acidic environments of bogs. Each small, round sundew leaf is covered in specialized glands that produce a sticky, dewlike substance. This glistening, sweet secretion not only attracts insects but also traps them on contact. The trapped insect triggers the sundew to produce digestive enzymes that dissolve the insect, allowing the plant to absorb essential nutrients such as nitrogen and phosphorus directly from its prey.

Hawley Bog Preserve, Hawley, MA
A classic peat bog maintained by the Nature Conservancy

Red Maple

Black Spruce

Balsam Fir

Highbush Blueberry

Bog Willow

Leatherleaf

Trees and Shrubs at the Edges of the Bog

Balsam Fir

Red Maple

There are typically many sundews embedded in the sphagnum moss, but the plants are *tiny*. Look carefully for them.

Round-Leaved Sundew

Grass Pinks

Black Spruce

Leather Leaf

Labrador Tea

Bog surface sphagnum moss layer

Lake water below the sphagnum surface layer

Old sphagnum peat moss covers the lake bottom

Lake bottom mud

Black Spruce, Red Maple, Leatherleaf, and
Highbush Blueberry along the bog edges

Black Spruce and
Labrador Tea

Pitcher Plant
flowers

Purple Pitcher
Plants

Plants at the Center of the Bog

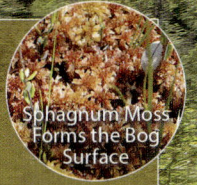

Sphagnum Moss
Forms the Bog
Surface

Purple Pitcher
Plant

Grass Pink

Round-Leaved
Sundew

Labrador Tea

Bog Laurel

Water depth below the
sphagnum layer can
vary from a few inches
to many feet.

There is a small area of open
water at Hawley Bog, but it is
far to the right of this view.

Purple Pitcher Plant
Sarracenia purpurea

Tucked away in the far north of New Hampshire, near the Canadian border, lies the Fourth Connecticut Lake. This small, pristine body of water is not just a scenic spot; it's the birthplace of the mighty Connecticut River. The lake is ringed by a floating bog mat, home to a variety of unique bog plants including sundews and pitcher plants, which have adapted to thrive in the nutrient-poor environment. The Fourth Lake is fed by runoff from the surrounding hillsides, and thus the water is not as acidic and nutrient-poor as a true bog.

Typical bog plants

The abundance of sphagnum moss in bogs is crucial. The moss retains water, maintains acidic conditions (low pH levels), and leaches nutrients from the surrounding environment, reinforcing the nutrient-poor conditions. The waterlogged and acidic conditions slow decomposition, accumulating peat as older mosses die and sink below the bog surface.

The most attention-grabbing feature of bogs is the presence of carnivorous plants such as the Purple Pitcher Plant and the Round-Leaved Sundew. Carnivorous plants trap and dissolve insects to supply crucial elements such as nitrogen, phosphorus, and potassium that are not found in the nutrient-poor soil. Pitcher plants use highly modified leaves to form pitchers of sweet liquid to entice flying and crawling insects into the pitcher, where a complex of hairs and slippery surfaces make it impossible for the insect to retreat. Insects become trapped in the sticky nectar of the pitcher and are gradually dissolved by enzymes and absorbed into the plant.

Sundews use hairy leaves tipped with nectar to attract and trap flying insects. While pitcher plants are easily spotted when you scan the surface of a bog, you might miss the thousands of sundews on the mossy surface because they are so small. A large, mature, Round-Leaved Sundew plant is typically only inches across and extends only a few inches above the surface of the sphagnum moss. Acid-loving wildflowers, small shrubs, and ferns also grow on the surface of bogs. Bog Rosemary, Royal Fern, Grass Pink, and the unusual but spectacular Dragon's Mouth orchid stand out against the background of sphagnum moss.

Small cold-adapted shrubs can also grow on the sphagnum moss surface of bogs. Plants such as Leatherleaf, Labrador Tea, and Small Cranberry are well adapted to thrive in acidic soil. Small Black Spruce trees can also grow on the mossy bog surface, although poor nutrients and a less-than-secure substrate usually combine to dwarf spruces that grow well into the bog itself. Black Spruce "saplings" are only a few feet tall and may be ancient, like natural bonsai trees dotting the bog surface.

The peripheral shorelines of bogs are rich with shrubs and trees that are tolerant of both acid and moist soils. An understory of shrubs including Leatherleaf, Labrador Tea, Highbush Blueberry, Bog Willow, Bog Laurel, and Sweetgale lines bog edges, often under a canopy of Black Spruce, Red Maple, Balsam Fir, American Larch (Tamarack), Red Spruce, and Northern White Cedar.

Bog animals

Bogs are not known for their rich animal and aquatic life because of the highly acidic and nutrient-poor conditions. Few fish and turtles can tolerate acidic water. A prominent exception is the tiny and quite rare Bog Turtle. Most amphibians are also scarce within the main body of bogs, but many frog and toad species can be found around the edges of bogs if the bog is fed by a stream with nonacidic water (*see illustration, pp. 196–97*). Bogs are richer in flying animals. Clouds of tiny flies and mosquitoes flying over and near bogs attract such birds as Barn Swallows and Tree Swallows, which specialize in capturing insects on the wing. The forests surrounding bogs are often home to typical wetland and moist forest songbirds such as the Common Yellowthroat and Palm Warbler.

The boglike edges of Northern ponds

Although true bog environments are rare even in the forests of northern New England, boglike environments dominated by sphagnum mosses and acid-loving plants are not uncommon. Boglike communities dominate the edges of many Northern ponds. Excellent examples of these habitats can be found in the spectacular Pondicherry National Wildlife Refuge in New Hampshire (*see illustration, pp. 198–99*).

The Pondicherry National Wildlife Refuge sits between the Connecticut River Valley and the White Mountains. Along with its many easily accessible examples of Northern wetlands and bogs, as well as outstanding Black Spruce swamps and lowland spruce-fir forests, the wildlife refuge contains two large but very shallow ponds. Botanists classify boglike edges of Northern lakes as poor fens—boglike wetlands that are fed at least in part by a stream. The steady flow of stream water reduces the acidity of the water and steadily feeds nutrients into the pond system. At Pondicherry, both Cherry Pond and

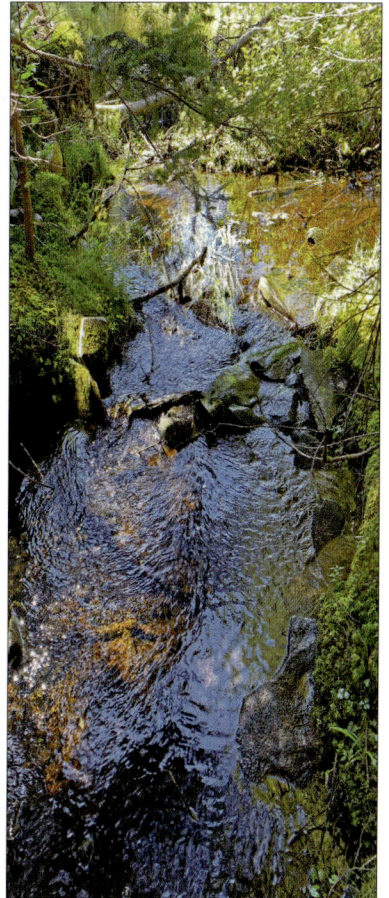

This small trickle of water running through the boggy mosses and sphagnum beds of a Black Spruce swamp is the origin stream of the Connecticut River as it emerges from New Hampshire's Fourth Connecticut Lake. Like many northern lakes, the Fourth Connecticut Lake is surrounded by swamps dominated by Black Spruce mixed with Red Spruce, American Larch, and Balsam Fir. The wet ground is thickly covered with sphagnum, Haircap, and other mosses and Cinnamon Ferns.

Potash Brook, at the edge of the Hawley Bog
A transition between the forest creek and bog environments

Black Spruce

Hawley Bog in the background

Red Maple

Red Maple

Black Spruce

American Larch

Tussock Sedge

Potash Brook

Spicebush Swallowtail

Balsam Fir

Royal Fern

Highbush Blueberry

Leatherleaf

American Bullfrog (female)

Bog Turtle

Yellow Pond-Lily

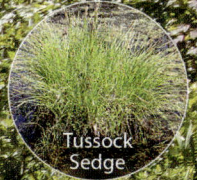

Tussock Sedge

Leatherleaf

Arrowwood Viburnum

Cinnamon Fern

Ebony Jewelwing

Black Chokeberry

Red Maple

Red Maple

Arrowwood Viburnum
Bog Laurel
Black Chokeberry
Leatherleaf

Leatherleaf
Highbush Blueberry
Bog Laurel
Royal Fern

Common
Cattail

Tussock Sedge

Eastern Pondhawk
Dragonfly (female)

Yellow Pond-Lily

Green Darner
Dragonfly

Animals are not shown to scale

Cherry Pond, part of the Pondicherry Wildlife Refuge, in the shadow of New Hampshire's White Mountains. Pondicherry is a significant part of the Silvio O. Conte National Fish and Wildlife Refuge of protected wildlife habitats throughout the Connecticut River watershed. Pondicherry has been called "one of the crown jewels" of New Hampshire's landscape, and it is an important refuge for rare plant environments and wild animals of all kinds. The boreal forests, bogs, fens, swamps, marshes, ponds, and grasslands support a wide variety of bird life. Cherry Pond is critical as a migration way station for many kinds of waterfowl, and the surrounding forests are full of songbird migrants in spring and fall.

Here a lowland spruce-fir swamp comes right up to the pond edge.

The pondside vegetation here is a Leatherleaf–Sheep Laurel–dwarf shrub bog, which also includes Labrador Tea, Bog Laurel, and Small Cranberry and Highbush Blueberry shrubs.

This area is a short emergent marsh dominated by Tussock Sedges, various bur-reeds, and other sedges. Mixed in area are patches of sphagnum bog, with Labrador Tea, Leatherleaf, and even scattered pitcher plants.

Plants found in or near Northern bogs and other highly acidic wetlands.

RED MAPLE *Acer rubrum*

BLACK SPRUCE *Picea mariana*

RED SPRUCE *Picea rubens*

BALSAM FIR *Abies balsamea*

AMERICAN LARCH *Larix laricina*

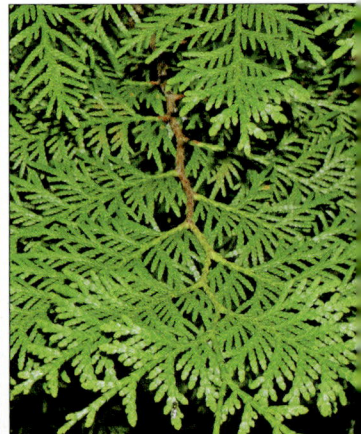

NORTHERN WHITE CEDAR *Thuja occiden*

BOG WILLOW *Salix pedicellaris*

HIGHBUSH BLUEBERRY *Vaccinium corymbosum*

LEATHERLEAF *Chamaedaphne calyc*

OG LAUREL *Kalmia polifolia*

LABRADOR TEA *Rhododendron groenlandicum*

SWEETGALE *Myrica gale*

MALL CRANBERRY *Vaccinium oxycoccos*

BOG ROSEMARY *Andromeda polifolia*

CREEPING SNOWBERRY *Gaultheria hispidula*

OYAL FERN *Osmunda regalis*

SPOONLEAF SUNDEW *Drosera intermedia*

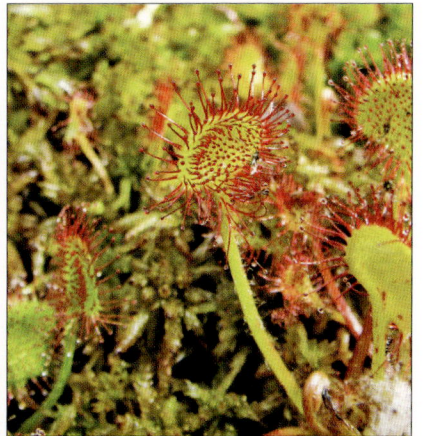

ROUND-LEAVED SUNDEW *D. rotundifolia*

Northern bog (top two rows, left page) and Southern pocosin plants.

DRAGON'S MOUTH *Arethusa bulbosa*

SWAMP LOOSESTRIFE *D. verticillatus*

ROSE POGONIA *Pogonia ophioglossoic*

STARFLOWER *Trientalis borealis*

PURPLE PITCHER PLANT *Sarracenia purpurea*

GRASS PINK *Calopogon tuberosus*

YELLOW PITCHER PLANT *S. flava*

YELLOW PITCHER PLANT FLOWERS

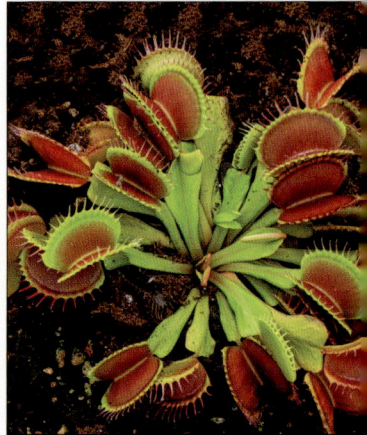

VENUS FLYTRAP *Dionaea muscipula*

Southern pocosin plants.

ND PINE *Pinus serotina*

POND PINE CONES *Pinus serotina*

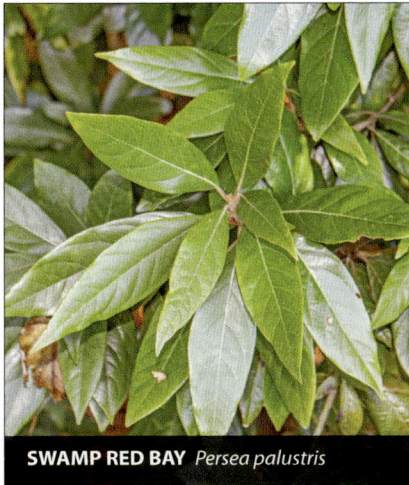

SWAMP RED BAY *Persea palustris*

LACK TUPELO *Nyssa sylvatica*

PROTHONOTARY WARBLER *P. citrea*

RED-COCKADED WOODPECKER *L. borealis*

ACCOON *Procyon lotor*

BLACK BEAR *Ursus americanus*

RED WOLF *Canis rufus*

Black Spruce
Picea mariana

Red Spruce
Picea rubens

American Larch (Tamarack)
Larix laricina

Sphagnum moss
Sphagnum sp.

Little Cherry Pond are supplied with water by the small Johns River. Both ponds are no more than six feet deep, and they contain rooted aquatic plants across most of their surface area. So, although Cherry Pond is more than 100 acres in size, it is a true pond by ecological definition. Over 200 species of birds—including more than 50 water-bird and shorebird species—have been identified at the Pondicherry Refuge, and Cherry Pond is one of New England's most important migration way stations for ducks, geese and other waterfowl heading south in autumn migration.

Moose, River Otters, Muskrats, and North American Beavers are active within the ponds and wetlands of Pondicherry. As a stream-fed and therefore less acidic poor fen environment, the ponds also host a wider variety of amphibians and reptiles than are found in a true bog. Most Northeastern frog and toad species occur in the refuge, Painted and Snapping Turtles are common, and the park is famed for its wide variety of dragonfly and butterfly species.

The Black Spruce swamps that surround the Pondicherry ponds are a cross between a sphagnum bog and a forested swamp. In many respects, they are the more northerly equivalent to Atlantic White Cedar swamps (*see illustration, p. 161*), and Atlantic White Cedar is also a common tree in Black Spruce swamps.

Black Spruce swamps are widely distributed across northern New England and southern Canada. They are characterized by their water-logged, acidic, and nutrient-poor soils, which support a distinct community of vegetation adapted to these harsh conditions. Under a thick canopy of Black Spruce, Red Spruce, and American Larch, the ground layer is typically blanketed by a dense mat of sphagnum and other forest mosses. The thick mosses are critical in maintaining the high water content and acidity of the swamp. This moss layer not only influences the soil conditions but also contributes to preserving the low nutrient levels by limiting decomposition—the typical communities of decomposing bacteria and fungi can't survive in the harshly acidic water. Interspersed among the mosses are various acid-loving shrubs such as Labrador Tea, Bog Rosemary, Lowbush and Highbush Blueberries, Leatherleaf, and carnivorous pitcher plants, which supplement their nutrient intake by trapping and digesting insects.

The Johns River, which feeds into Cherry Pond at the Pondicherry Wildlife Refuge in Whitefield, New Hampshire. As it enters Cherry Pond, the Johns is really more of a boggy creek that emerges from a larger marsh area to the west of Cherry Pond. Ecologists class the marshes at Pondicherry as an interesting mix of poor bog-fen vegetation, blended into what is mostly a classic wet meadow that surround both the Johns River and Cherry Pond (*see illustration, pp. 198–99*). Along with the common Tussock Sedges, Cinnamon and Royal Ferns, and Highbush Blueberries, you might spot boglike patches of sphagnum moss, Purple Pitcher Plants, and (in June) the bright flowers of Grass Pinks.

American Sweetgum
Liquidambar styraciflua

Inkberry
Ilex glabra

Redbay
Persea borbonia

Pocosin wetlands and ponds

Pocosin ponds, also known as Carolina bays, are unique and ecologically significant wetland ecosystems found along the Atlantic Coastal Plain, primarily in North Carolina and Virginia, but extending south along the coastal plain to northern Florida. On the Delmarva Peninsula, similar shallow, acidic ponds and swamps are called Delmarva bays or, locally, whale wallows. About 70 percent of pocosin wetlands are located on the coastal plain of North Carolina. Sandy rims, acidic waters, and a unique array of plant and animal life characterize these shallow wetlands. Pocosins are similar to northern bogs in that these acidic wetlands on a thick bed of organic matter are surrounded and dominated by evergreen tree species and plants and shrubs adapted to the nutrient-poor and acidic water environment.

The term "pocosin" is derived from the Algonquian word meaning "swamp-on-a-hill," which accurately describes the slightly elevated, swampy terrain where these raised bogs are located. This unique positioning contributes to the pocosins' isolation, making them relatively nutrient-poor, acidic environments due to the accumulation of organic matter in peat. As in Northern bogs, pocosins accumulate peat because the acidic water inhibits the presence of the usual communities of bacteria and fungi that break down vegetal organic matter. Despite these limiting conditions, pocosin ponds support a rich biodiversity and serve as crucial breeding grounds and habitats for numerous amphibians, reptiles, birds, and invertebrates.

The vegetation surrounding and within pocosin ponds is uniquely adapted to the acidic, water-logged, and nutrient-poor conditions. Carnivorous plants such as the Venus Flytrap and Yellow Pitcher Plants are iconic residents. These plants use their bug-trapping features to supplement their intake of nutrients, since the sandy soils in pocosin areas are often poor and leached of minerals.

The vegetation of pocosin ponds is dominated by the Pond Pine, a fire-adapted tree species that thrives in nutrient-poor soils. Other common plant species include Fetterbush, Titi, Whortleberry, Bilberry, Cranberry, Labrador Tea, and various sedges and grasses. The vegetation surrounding pocosin ponds often transitions into pocosin wetlands, characterized by dense shrub swamp thickets and a high water table.

Pocosin ponds play a significant role in the hydrology of the Mid-Atlantic Coastal region. They are natural reservoirs, capturing rainfall and slowly releasing water, thus moderating streamflow and reducing the risk of flooding in surrounding areas. In addition, the peat soils of the pocosins are significant carbon sinks, storing vast amounts of carbon that, if released, would contribute to greenhouse gas emissions. This aspect of pocosin pond ecology underscores the importance of

The Red Wolf (*Canis rufus*), once declared extinct in the wild, remains one of the most endangered canids in the world. Coastal North Carolina hosts the only wild population of these wolves, centered in the Alligator River National Wildlife Refuge and surrounding areas. After successful reintroduction efforts in the 1980s by the U.S. Fish and Wildlife Service, the population initially grew but since has faced significant challenges. Recent estimates suggest that fewer than 40 wolves remain in the wild due to threats from habitat loss, vehicle strikes, and illegal killings. Efforts to stabilize the population include adaptive management techniques such as fostering captive-born pups into wild packs and implementing sterilization programs for coyotes to prevent hybridization. Continuous monitoring and community engagement remain critical to the recovery of the Red Wolf in its native habitat.

Pocosin Wetland, Alligator River NWR, North Carolina
Unique acidic wetlands of the Southeastern Coastal Region

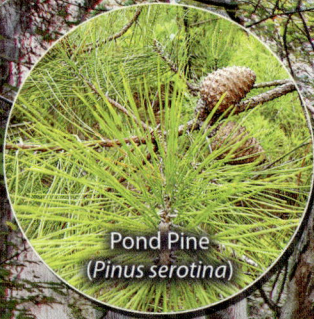

Pond Pine
(*Pinus serotina*)

Swamp Bay
(*Persea palustris*)

Venus Flytrap
(*Dionaea muscipula*)

Yellow Pitcher Plant
(*Sarracenia flava*)

Southern
Five-Lined Skink
(*Plestiodon inexpectatus*)

American
Green Treefrog
(*Hyla cinerea*)

PROTHONOTARY
WARBLER

SWAINSON'S
WARBLER

WOOD DUCK
(Female)

Wood Duck ducklings

North Carolina's Lake Mattamuskeet is a crucial habitat for various migratory birds. Renowned for its abundant wildlife, Lake Mattamuskeet hosts an impressive array of migratory birds, including thousands of Tundra Swans, Snow Geese, and various duck species. The surrounding wetlands and pocosin forests further enrich its biodiversity, providing habitat for Black Bears, Bobcats, River Otters, and the highly endangered Red Wolf.

pocosins in mitigating the effects of climate change. Unfortunately, the pocosin wetlands were not valued or protected in the past, and we've lost an estimated one-third of pocosin wetlands during the past century.

Lake Mattamuskeet is North Carolina's largest natural lake, covering 40,000 acres. It lies within the expansive Lake Mattamuskeet National Wildlife Refuge, which spans over 50,000 acres. The lake and surrounding refuge are one of the most heavily used natural habitats for migrating and wintering waterfowl on the East Coast.

Mattamuskeet is massive but shallow, with an average depth of only two to three feet. This allows sunlight to penetrate the bottom, fostering an abundant surface and submerged aquatic plant life. This plant life forms the foundation of the lake's ecosystem, supporting a complex food web. Dominant aquatic plants include various pondweeds, water lilies, and American Lotus, which provide food and shelter for numerous aquatic animal species.

The surrounding wildlife refuge boasts a mixture of habitats, including marshes, wooded swamps, and open waters, which contribute to its rich biodiversity. It also includes many wetlands that are very similar to pocosins in most respects. The lake and its environs are particularly renowned as a wintering site for large flocks of Tundra Swans, Snow Geese, and a variety of duck species, making it a prime location for birdwatching. The refuge also supports various raptors, including Bald Eagles, Peregrine Falcons, and Ospreys.

Lake Mattamuskeet is inhabited by a range of freshwater fish species, such as Largemouth Bass, Bluegill, and American Eel, which attract River Otters and various wading birds such as herons and egrets. The lake's margins, with their moist soil, support a thick growth of cattails and bulrushes, which are crucial avian nesting and feeding grounds.

The combination of open water, marsh, and woodland also supports such mammals as White-Tailed Deer, Raccoons, and the occasional Black Bear. This region was also the core territory of the highly endangered Red Wolf, locally extirpated by the 1960s. Red Wolf reintroduction efforts by the U.S. Fish and Wildlife Service have been hampered by local opposition and illegal killings and trappings of Red Wolves.

Bluegill
Lepomis macrochirus

Where to see pocosins and related wetlands

With their dense tangle of vegetation, very soggy soils, and large areas of standing water of uncertain depth, pocosin habitats can be challenging to explore. The Pocosin Lakes National Wildlife Refuge and the Alligator River National Wildlife Refuge, both just inland from North Carolina's Outer Banks, are excellent locations for exploring pocosin wetlands. There are numerous raised roads through the reserves, and you can quickly stop and watch for birds or check out the vegetation without getting soaked. The nearby Lake Mattamuskeet National Wildlife Refuge is also well worth exploring. Although the lake is not a pocosin, its shores have very similar plants and animals, and the refuge's roads also offer opportunities to explore pocosin habitats without the need for serious swamp bushwhacking.

The Alligator River National Wildlife Refuge in coastal North Carolina spans over 152,000 acres of diverse habitats, including wetlands, hardwood swamps, and marshes. Established in 1984 to preserve and protect unique pocosin and other wetland habitats, the refuge supports a wide variety of wildlife, including the endangered Red Wolf, and is one of the northernmost habitats of the American Alligator. The refuge is also a haven for migratory birds and, through its extensive road system, offers access to pocosin wetland wild areas that are otherwise quite difficult to visit.

All kettle ponds originated as giant chunks of ice left behind by the Wisconsinan Glacial Period (the Ice Age). Most surviving kettle ponds are along the New England coast, but some kettles lie deep in New England interior areas that were deeply covered by the Ice Age glacier. This is Kettle Pond, in Vermont's Kettle Pond State Park, viewed from the peak of nearby Owls Head Mountain.

KETTLE PONDS

Kettle ponds, sometimes called kettle holes, are ponds formed by glacial ice. In the most recent glacial period, the Wisconsinan Glacial Episode, or Ice Age, the melting of the massive Laurentide Ice Sheet left behind giant chunks of ice, some the size of icebergs. These vast ice blocks, usually at least partly buried within sandy glacial till, eventually melted, leaving ponds scattered across the postglacial landscape. Kettle ponds are typically deep for their size and often have smooth, rounded shorelines, leading to the comparison of their shapes to kettles (*see illustrations, p. 218*).

Visitors to the southeastern coastal region of New England are often surprised by its many lakes and ponds, particularly on Cape Cod, but also on eastern Long Island and the islands of Nantucket, Martha's Vineyard, and Block Island. This very sandy, glaciated region of southeastern New England is called the Outer Lands (*see illustration, overleaf*) because of its geological origins and the unique communities of plants and animals living on its sandy soils. It's hardly intuitive that the mostly sandy and glacial till remains of the last Ice Age would have such an abundant freshwater supply, yet a crucial coupling of chemistry and geology makes freshwater possible and common in the Outer Lands. Freshwater is less dense than salt water, and at the scale of large groundwater bodies, freshwater floats above the salt water that permeates the deep layers of glacial sediment under Cape Cod and the nearby islands. Although the sandy sediments that make up the Outer Lands are very porous, the enormous volumes of sandy earth act like a giant sponge, preventing rainwater and snowmelt from simply running off the land and into the sea.

The glacial origin of kettle ponds

There is no formal lexicographic definition of a pond versus a lake. By local tradition, even the largest freshwater bodies on Cape Cod

The nation's best-known kettle pond isn't on Cape Cod or near the coast. In 1845, Henry David Thoreau retreated to a small cabin near Walden Pond on the outskirts of Boston, Massachusetts. "I went to the woods because I wished to live deliberately, to front only the essential facts of life, and see if I could not learn what it had to teach, and not, when I came to die, discover that I had not lived." Today, Walden Pond is surrounded by the busy suburb of Concord, but thanks to the Walden Pond State Reservation, the pond retains some of the contemplative beauty Thoreau once knew. Unfortunately, the proposed expansion of a local airport and a nearby large housing development will challenge Thoreau's beloved pond.

Boston

MASSACHUSETTS

Provincetown

*Cape Cod
Bay*

RHODE
ISLAND

Cape Cod

CONNECTICUT

*Nantucket
Sound*

NEW
YORK

New Haven

*Martha's
Vineyard*

Nantucket

*Block Island
Sound*

*Block
Island*

Long Island Sound

The Outer Lands

Port Jefferson

Long Island

are usually called ponds, probably because most originated as kettle ponds. When the glaciers melted out of the Cape Cod area about 20,000 years ago, they left behind iceberg-sized chunks of ice buried in the glacial till that later melted to form ponds. There are a lot of kettle ponds—depending on how you count them, Cape Cod has more than 500 of all sizes—and from the air, the Cape landscape can sometimes appear half-liquid with the reflections from these hundreds of ponds. Cape Cod has many more kettle ponds than the islands because large sections of the southern outwash plains of Martha's Vineyard, Nantucket, and Long Island were never covered by glacial ice, and thus, there were fewer blocks of remnant ice to create surface ponds. The glacial ice sheet covered the entire land surface of what is now Cape Cod and southwestern coastal Rhode Island, and the glacier left behind thousands of large ice chunks on both the upland moraine areas and the southern outwash plains of the Cape Cod landscape.

Kettle ponds are also found in mainland New England, Upstate New York, and all other places covered by the Laurentide Ice Sheet as recently as 25,000 years ago. Though not as numerous as the kettle ponds in sandy coastal regions, inland kettles were formed the same way: as the Ice Age glaciers melted, they often left behind gigantic chunks of ice half-buried in the glacial till, and those ice chunks eventually melted into ponds. Most mainland kettle ponds formed by the Ice Age silted up and disappeared, but small streams often feed the kettle ponds that survive today. Some of these inland kettle ponds are well known. Thoreau's Walden Pond in Concord, Massachusetts, is a kettle pond. Chocorua Lake in New Hampshire's White Mountains is a large kettle pond, and Vermont has Kettle Pond State Park, which contains about a dozen smaller kettle ponds. In the middle of New York's Long Island, Lake Ronkonkoma is a large kettle pond over 65 feet deep, very unusual for a sandy region that mostly has small, shallow freshwater ponds.

Why kettle ponds are often round

Kettle ponds differ from typical ponds on the mainland in several important ways. As ponds derived from gigantic chunks of ice buried in glacial till, most kettle ponds are unusually deep for their size, with a kettle-like cross-section (*see illustration, p. 218*). The name "kettle" also refers to the rounded shorelines of most kettle ponds. Today, most kettle pond shorelines are smooth arcs not because the original glacial ice was rounded but because, over thousands of years, the natural cycles of wind and water movement have eroded and smoothed the soft earthen pond shores, much as ocean waves and longshore currents smooth ocean beach profiles.

Kettle ponds in the Outer Lands are also unusual because, unlike as in most mainland ponds, substantial surface streams do not typically

From space, Cape Cod is so pockmarked with kettle ponds that it looks like Swiss cheese. Even many of the coastal bays and inlets are actually old kettle ponds that have since been flooded with salt water. All the ponds originated as gigantic chunks of ice, left behind as the Wisconsinan Glacial Episode (the Ice Age) ended about 12,000 years ago.

Little Cliff Pond, in Nickerson State Park on Cape Cod, owes its existence to the region's glacial past. During the Ice Age, retreating glaciers left scattered giant ice blocks half-buried in the sandy landscape. Over hundreds of years these glacial ice blocks melted in the warming environment to form kettle ponds like Little Cliff Pond. The surrounding area is characterized by sandy, well-drained soil typical of Cape Cod's glacial outwash plains. The pond's ecosystem is a unique blend of freshwater and coastal influences, supporting a variety of aquatic plants and fish species. The clear waters of Little Cliff Pond are currently threatened by groundwater pollution from nearby commercial and residential development. As nitrogen, phosphorus, and other nutrients and pollutants seep into the groundwater and enter the pond, the excess nutrients cause harmful algal blooms that limit or even kill the normal pond inhabitants. Although Nickerson State Park is well maintained, coastal freshwater environments cannot be shielded from the surrounding developed environment.

Kettle ponds (*see below*) differ from typical ponds in several important ways. As ponds derived from gigantic chunks of ice buried in glacial till, most kettle ponds are unusually deep for their size, with a kettle-like cross-section. The name "kettle" also refers to the rounded shorelines of most kettle ponds. Today, most kettle pond shorelines are smooth arcs not because the original glacial ice was rounded but because, over thousands of years, the natural cycles of wind and water movement have eroded and smoothed the soft earthen pond shores, much as ocean waves and longshore currents smooth ocean beach profiles.

feed them. The land area on eastern Long Island, the islands, and Cape Cod is too small and the soils are too porous to have upland drainage areas supporting substantial surface streams or rivers. Kettle ponds derive their water almost entirely from groundwater sources, either because the pond is fed by underground springs or because the pond surface intersects the local groundwater level.

Acidity and pollution

The sandy soils of the coastal regions don't contain the wide range of natural minerals that help buffer mainland aquatic systems. Hence, the water of kettle ponds is often more acidic than mainland ponds, shifting the profile of aquatic organisms and plants to those that withstand higher acidity. Pitch Pines, Loblolly Pines, Slash Pines, and other conifers have surrounded many coastal ponds for thousands of years, and the conifer needles are yet another source of natural acids.

Aerial view of kettle ponds in Wellfleet, on Cape Cod, Massachusetts.

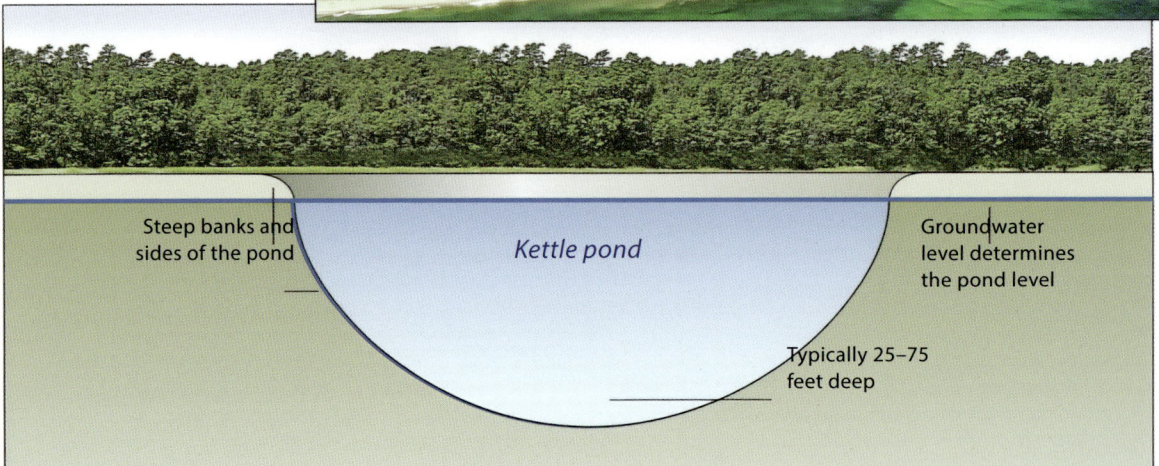

Steep banks and sides of the pond

Kettle pond

Groundwater level determines the pond level

Typically 25–75 feet deep

The smooth shorelines of Walden Pond and other kettle ponds are due to at least 15,000 years of wind and wave erosion. Although most kettle ponds are small, they can be unusually deep and steep-sided below the surface, with not much emergent vegetation along their shores.

This acidity, combined with an ordinarily scarce nutrient profile, means that coastal ponds are susceptible to pollution and eutrophication caused by excess nitrogen and phosphorus in runoff from sewage systems, home septic fields, and wastewater that contains soap residues. Many coastal ponds turn a deep green in the warmer months because of excess algae growth. In very polluted ponds, there may also be fish kills and deaths of other aquatic animals due to hypoxia—a very low level of dissolved oxygen in the water. Excess nutrients enter the ponds, causing explosive growth of usually scarce algae, and the algae use up the dissolved oxygen in the water at night when there is no photosynthesis. The process of eutrophication and hypoxia in ponds is very similar to what happens in coastal marine waters. Unlike in marine systems, ponds have no daily tidal flushing, so even a tiny amount of excess nutrients can cause long-term damage to pond ecosystems.

Herring runs and kettle ponds

Many coastal regions, including Cape Cod, Martha's Vineyard, Nantucket, Long Island, and all barrier islands, do not have enough drainage area to support significant rivers. This lack of rivers is a challenge to anadromous fish species that run upstream in the spring to lay eggs. Coastal and kettle ponds play an important role in offering freshwater habitat in regions that lack river significant habitat. Most kettle ponds have an exit stream that carries excess water out of the pond and down to sea level. In precolonial times, these pond exit

Alewives (*Alosa pseudoharengus*) are anadromous fish belonging to the herring family. They spend most of their adult lives in the ocean but return to freshwater environments to spawn. These fish typically grow to 10–12 inches in length and have a distinctive silvery appearance with a dark spot behind the gills. Alewives' reliance on small coastal streams makes this species particularly vulnerable to habitat degradation and barriers to migration, highlighting the importance of preserving connectivity between marine and freshwater ecosystems.

streams played a crucial role in the life cycles of anadromous fish, which breed in freshwater ponds but spend most of their adult lives in the ocean. Each spring, so-called river herrings such as the American Shad, Alewife, and Blueback Herring would swim up the exit streams to lay their eggs in the kettle ponds. Coastal streams also once supported the unusual catadromous life cycle of the American Eel, which lives in coastal waters, freshwater lakes, and streams but breeds deep in the central Atlantic Ocean.

In colonial and early American times, these small coastal streams were ideal for building water-powered mills and small factories, and many streams were dammed, ending the fish runs. Unfortunately, many of those old mill dams remain in place, centuries after the mills rotted away and the dams became useless. In recent years, states and river advocacy groups have begun removing or at least breaking open hundreds of old dams. In these coastal areas "upriver" usually ends in a kettle pond or coastal pond. Restoring the anadromous fish runs is critical for preserving Striped Bass, Atlantic Salmon, Alewives, American Eels, and Blueback Herrings. The millions of migrating fish were once an important nutrient source for inland areas near ponds and streams. The fish eggs and the bodies of breeding fish that died or were eaten by birds and other predators along the journey were like a giant regional nutrient stream, carrying the riches of coastal waters far inland. The restoration of herring runs reaps benefits far beyond the recovery of anadromous fish species—it enriches all coastal wild environments.

A number of streams and kettle ponds on Cape Cod are famous for their herring runs in early spring. The runs up small Cape streams are primarily composed of Alewives but also include Blueback Herring and American Shad. The fish run upstream to breed as soon as the air temperature rises above 50°F, as this gives the eggs a chance to mature before sunfish and other egg-eating pond fish become active. The springtime start of these herring runs is usually signaled by masses of gulls that crowd the stream banks and airspace above the stream, feasting on the migrating fish in the shallow streams.

Here, Herring Gulls pluck Alewives directly from the shallow Stony Brook in Brewster, on Cape Cod.

Common plant species around coastal plain freshwater ponds and coastal kettle ponds.

BEACH PLUM *Prunus maritima*

BLACK CHERRY *Prunus serotina*

SASSAFRAS *Sassafras albidum*

QUAKING ASPEN *Populus tremuloides*

STAGHORN SUMAC *Rhus typhina*

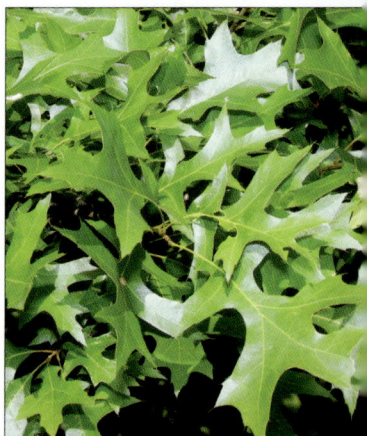

NORTHERN RED OAK *Quercus rubra*

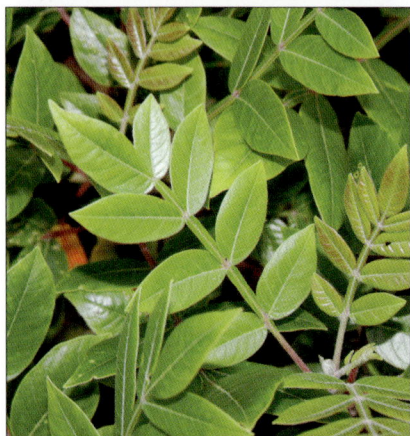

SHINING (WINGED) SUMAC *Rhus copallina*

BEAR OAK *Quercus ilicifolia*

AMERICAN HOLLY *Ilex opaca*

BLACK OAK *Quercus velutina*

WHITE OAK *Quercus alba*

RED MAPLE *Acer rubrum*

POISON IVY *Toxicodendron radicans*

VIRGINIA CREEPER *Parthenocissus quinquefolia*

CATBRIER *Smilax glauca*

NORTHERN BAYBERRY *Myrica pensylvanica*

HIGHBUSH BLUEBERRY *Vaccinium corymbosum*

EASTERN REDCEDAR *Juniperus virginiana*

This old kettle pond in Wellfleet on Cape Cod has been modified with a small dam that expanded and deepened the pond to fill a small valley to the left. The broader, less steep-sided pond now supports a rich and varied freshwater community, including small marshes and a rim of emergent vegetation.

The rich coastal ponds and wetlands of New Jersey's Cape May Point State Park seem contradictory: How could freshwater environments like this exist within sight of beaches and the Atlantic Ocean? And yet, sandy peninsulas and barrier islands often host freshwater ponds and marshes.

COASTAL PONDS AND WETLANDS

The Atlantic Coastal Plain, stretching from New Jersey to Florida, has a unique and often overlooked ecological feature: freshwater or slightly brackish ponds on or near the shoreline and barrier islands. Coastal plain ponds are also common on the sandy southeastern shores of New England and Long Island.

The origins of these ponds are deeply intertwined with the geological history of the Atlantic Coastal Plain. In areas north of New York Harbor, most coastal ponds are relics of the Ice Age, formed when retreating glaciers left behind depressions in glacial till and sand filled with freshwater. As the climate warmed and sea levels rose, these depressions became the foundation for the coastal ponds we see today. Farther south, ponds have formed more recently as the dynamic processes of barrier island migration created new habitats for pond development.

These ponds represent a convergence of freshwater and saltwater ecosystems. The proximity of these ponds to the ocean or to significant estuaries creates a distinctive mix of freshwater and estuarine brackishness. Although many are primarily freshwater ponds, tides and the potential influx of salt water during nor'easters and hurricanes are influences. Also, all plants within at least five miles of an ocean coast are affected by salt aerosols, or sea spray (*see illustrations, overleaf*). This mixing of coastal and inland water types is a critical factor in shaping the ecology and biology of these coastal pond ecosystems.

Ecologically, these ponds serve as biodiversity hotspots, providing critical habitat for many plants and animals. The pond edges are often fringed with salt-tolerant grasses and sedges, creating a transitional zone between the aquatic and terrestrial environments. This vegetation is crucial in stabilizing the pond edges and provides shelter for various organisms. Submerged aquatic plants thrive within the pond,

Lighthouse Pond at New Jersey's Cape May Point State Park, with the famous Cape May Lighthouse in the background. Cape May's freshwater environments are protected from inundation by the surrounding ocean and bay by extensive dune fields. These are true freshwater ponds, and they host Largemouth Bass, various sunfish, Bullfrogs, Northern Watersnakes, Painted Turtles, and Snapping Turtles. Cape May's ponds and wetlands are also critical habitat for the tens of millions of birds that migrate along the Atlantic Coast in the spring and fall.

Mummichog
Fundulus heteroclitus

Banded Killifish
Fundulus diaphanus

offering food and habitat for a diverse community of invertebrates, including insects, crustaceans, and mollusks.

The fauna of these coastal ponds is equally diverse and specially adapted to the unique conditions of these habitats. Small but numerous fish species such as Mummichogs and Banded Killifish thrive in these coastal ponds and wetlands. These fish species have evolved physiological adaptations that allow them to tolerate fluctuating salinity levels. These smaller fish often become prey for larger predators, including wading birds such as herons and egrets and diving birds such as cormorants.

Amphibians also play a significant role in the ecology of these ponds, with various species of frogs and salamanders finding suitable habitats for breeding and larval development. The life cycles of these amphibians are intricately linked to the availability of freshwater, making these coastal ponds critical for their survival in the broader coastal ecosystem. The presence of amphibians in these ponds also indicates environmental health, as many amphibian species are sensitive to changes in water quality and habitat conditions.

A significant challenge faced by the inhabitants of these coastal ponds is the threat of desiccation. During drought or extreme heat, coastal ponds on sandy sediments may shrink significantly or even dry up. To survive these harsh conditions, some organisms have developed remarkable strategies. Coastal plant communities are often dominated by annual species with seeds that can remain dormant in the sand or soil for extended periods, awaiting the return of favorable conditions.

Salt aerosols shape environments well inland from the coast

Churning ocean surf full of air bubbles

Onshore winds ➡️

Stream of salt aerosols

Microscopic salt droplets from bursting bubbles form an invisible aerosol

Near the surf and beach the salt coating on plants is heavy and fatal to all but a few plants such as Sea Rocket and American Beach Grass.

Ocean coast   Wet beach   Upper beach

Salt aerosols from the ocean

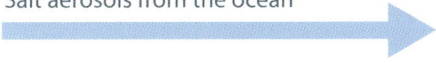

Exposed twigs and buds die from exposure to salt spray

More buds and branches die

"Windswept" appearance from salt pruning

Coastal trees are not windswept, they are salt-pruned.

The effects of salt aerosols can reach inland for miles

Even trees and plants well inland can be stunted and shaped by salt aerosols

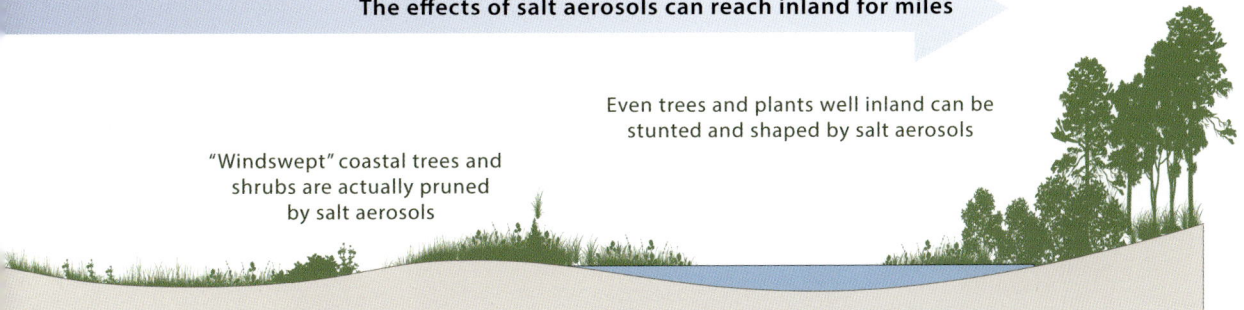

"Windswept" coastal trees and shrubs are actually pruned by salt aerosols

Dune grasslands

Coastal ponds and wetlands

Maritime forest

Trustom Pond National Wildlife Refuge in coastal Rhode Island features a diverse freshwater and saltwater pond ecosystem. The varying salinities create unique habitats supporting a wide range of flora and fauna. The refuge's largest water body, Trustom Pond, is shown here. The large, shallow pond is a complex, seasonally changeable mix of fresh and brackish environments. In the rainier parts of the year, the pond is almost entirely fresh. In the drier summer and early autumn months, and when storm surges force beach waves into the pond, the water may be salty for weeks at a stretch.

Coastal ponds exist in a delicate balance of saltwater influx from the ocean and freshwater input from streams and groundwater. For example, virtually every aquatic and wetland plant visible in this photo is a typical freshwater species. Still, the occasionally brackish conditions inhibit many other freshwater plants that cannot tolerate even slightly brackish water. This dynamic environment supports migratory waterfowl, wading birds, and numerous fish species. The surrounding wetlands and uplands further contribute to the area's ecological complexity, providing crucial nesting and feeding grounds for various resident and migratory wildlife.

Pitch Pine
Pinus rigida

Northern Bayberry
Myrica pensylvanica

Some invertebrates can enter a state of cryptobiosis (a dormancy strategy), suspending their metabolic processes until water returns. Fish species such as killifish can bury themselves in the mud at the bottom of drying ponds, surviving in small pockets of moisture until conditions improve.

The dynamic nature of these coastal ponds also influences their community structure and ecological interactions. The potential for periodic disturbances, such as storm surges or saltwater intrusion, can reset the environmental succession of these ponds. This creates opportunities for pioneering species and helps maintain a diverse community of organisms adapted to different stages of pond development. The result is a mosaic of habitats within and among ponds, supporting various species and ecological niches. This mixed environment is often a deliberate strategy employed by coastal refuge managers to support mosquito control measures and to create attractive habits for a wide range of coastal migrating birds. For example, managers at both New Jersey's Edwin B. Forsythe National Wildlife Refuge and Florida's Merritt Island National Wildlife Refuge maintain extensive freshwater and brackish ponds and impoundments, using floodgates to control salinity levels in these coastal wetlands and ponds.

The nutrient dynamics in these coastal ponds are complex and play a crucial role in shaping their ecology. The interface between freshwater

and marine environments creates unique biogeochemical conditions, influencing the cycling of nutrients such as nitrogen and phosphorus. These ponds can sometimes act as nutrient sinks, trapping and processing excess nutrients that might otherwise flow into coastal waters. This function can mitigate the effects of nutrient pollution on nearby marine ecosystems. These coastal wetlands are also critical carbon sinks, absorbing and trapping billions of tons of carbon compounds and carbon dioxide that might otherwise enter the atmosphere and contribute to global warming.

Coastal ponds also play a vital role in the broader ecosystem, serving as important stopover sites for migratory birds. Many shorebirds, waterfowl, and songbirds rely on these ponds as resting and feeding areas during their long migrations. The abundant invertebrate life in these ponds provides crucial nutrition for these birds, helping them to build up the energy reserves needed for their journeys.

From a conservation perspective, coastal ponds face numerous challenges. Climate change and sea-level rise pose significant threats, potentially altering the delicate balance of freshwater and salt water that defines these ecosystems. Increased storm intensity and frequency can lead to more frequent saltwater intrusion events, potentially pushing some ponds beyond the tolerance limits of their freshwater-adapted species. For example, in 2012, most of the freshwater and brackish

Black Cherry
Prunus serotina

The quintessential coastal pond:
Cape Cod's Blackwater Pond is a complex of ponds, emergent marshes, and wet meadows 35 miles out to sea from mainland Massachusetts, at the northern tip of Cape Cod. The whole complex of the Blackwater wetlands and the surrounding Beech Forest area of the Cape Cod National Seashore exists entirely on sand, surrounded on three sides by ocean beaches rarely less than a mile distant. And yet, a rich freshwater community survives and thrives in a maritime environment.

New Jersey's Edwin B. Forsythe National Wildlife Refuge lies along New Jersey's Atlantic coastline just north of Atlantic City. In addition to vast salt marshes and brackish wetlands, Forsythe also includes a number of large freshwater coastal ponds and marshes. Here, Lily Lake is actually a vast shallow freshwater pond, full of all the freshwater plant and animal species you might find in any inland pond or marsh.

Surrounded by a lush maritime forest, the freshwater ponds and marshes of Buxton Woods lie on Hatteras Island, a barrier island 30 miles off the mainland coast of North Carolina. Buxton Woods is the northernmost limit for the Saw Palmetto, the nearly ubiquitous Southern coastal plant. Although their location is exotic, the freshwater areas of Buxton Woods contain all the usual wetland plants and animals you might see in mainland ponds and marshes.

Saw Palmetto
Serenoa repens

areas of the Prime Hook National Wildlife Refuge on Delaware Bay were inundated by salt water when Hurricane Sandy breached an outer barrier beach and filled the marshes with a salty storm surge from the bay. The saltwater flooding made an almost instant and radical change to the ecology of the Prime Hook marshes. In the dozen years since Sandy, managers at the refuge have repaired the barrier breach and restored the freshwater and brackish areas of the marshes. Human development along the coast also threatens these freshwater and brackish habitats—pollution, habitat fragmentation, groundwater pumping, and altered tidal flows all pose risks to the health and persistence of these unique ecosystems.

Despite these challenges, the coastal ponds of the Atlantic Coastal Plain have demonstrated remarkable resilience over thousands of years. They have persisted through significant changes in climate and sea level, adapting to shifting conditions in their environment. It's important to remember that when coastal habitats are described as delicate or storm-prone, it's not the wild areas that are most at risk; it's the human-made structures and roads that are delicate and vulnerable.

These coastal ponds are more than isolated bodies of water; they are integral to the broader coastal landscape. They serve as critical habitats for most of our major sport fish and as nurseries for commercially important fish and other seafood. Coastal wetlands filter pollutants from the watershed and provide vital ecosystem services that benefit both wild and human communities.

Southeastern coastal ponds and wetlands

Along the Southeastern coastline, expansive salt marshes dominate. These marshes are characterized by cord grasses and needle rushes that thrive in brackish conditions. The marshes serve as crucial nurseries for many fish species and provide foraging grounds for migrating waterfowl and millions of wading birds such as herons and egrets, as well as vast flocks of ducks, geese, and shorebirds.

Florida Scrub Jay
Aphelocoma coerulescens

The Merritt Island National Wildlife Refuge, just north of Florida's Kennedy Space Center, contains many excellent and accessible examples of Southeastern coastal wetlands in freshwater, brackish water, and salt water. Merritt Island NWR encompasses diverse ecosystems, each supporting a unique assemblage of flora and fauna. The refuge's landscape is a mosaic of coastal and inland habitats shaped by the interplay of freshwater and saltwater influences. In addition to all types of wetlands and coastal ponds, Merritt Island is an excellent place to find endangered coastal scrub and dune species such as the Florida Scrub Jay and the Gopher Tortoise.

Gopher Tortoise
Gopherus polyphemus

Merritt Island's extensive network of impoundments was originally created for mosquito control but is now managed to maximize its value for resident and migratory birds and preserve Florida's unique mix of coastal wetland animals and plant life. On the Florida coast,

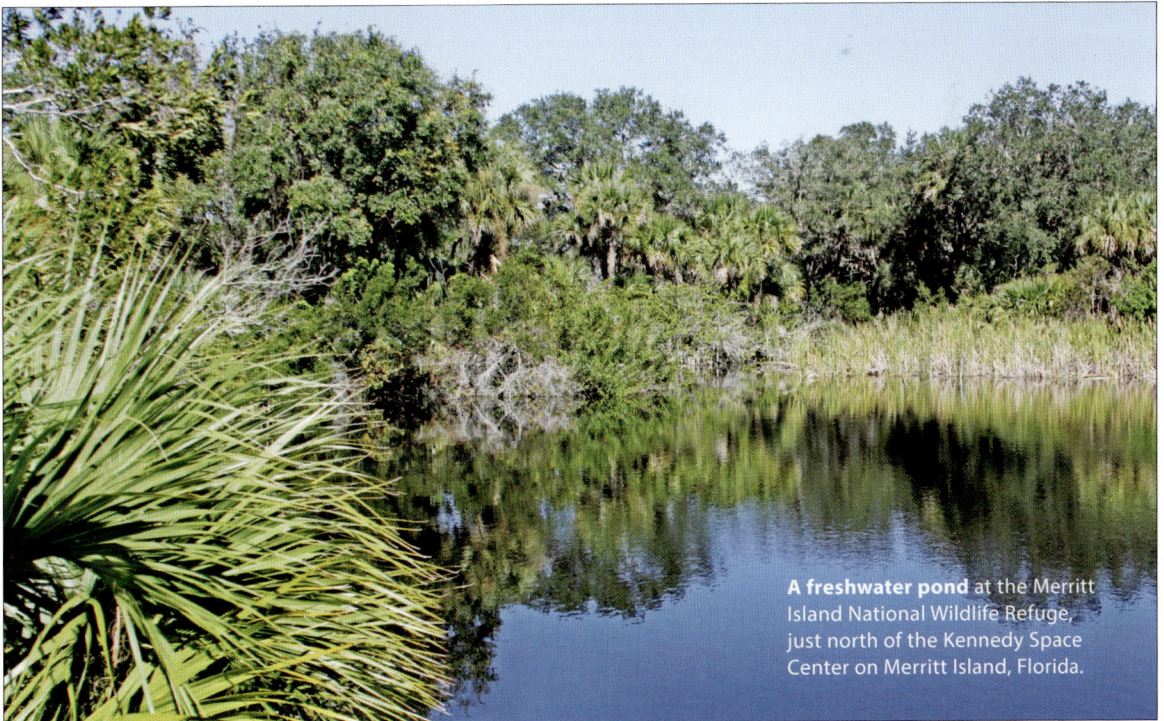

A freshwater pond at the Merritt Island National Wildlife Refuge, just north of the Kennedy Space Center on Merritt Island, Florida.

there are only a handful of significant refuges and parks to protect this complex and unique environment, which is now mostly destroyed and buried under mile after endless mile of coastal condominiums and apartment buildings, all barely above sea level.

The refuge's brackish lagoon systems, including portions of the Indian River Lagoon and Mosquito Lagoon, represent some of North America's most biologically diverse estuaries. These waters support abundant marine life, from microscopic plankton to charismatic West Indian Manatees and Bottlenose Dolphins. The lagoons' seagrass beds are vital, offering shelter and sustenance for many estuary and marine species. These shallow water bodies vary in salinity and depth, creating a range of microhabitats. Some impoundments are kept predominantly as freshwater, attracting migratory waterfowl and wading birds. Others can fluctuate in salinity, mimicking natural coastal wetlands and supporting a mix of freshwater and estuarine species.

The biology of these Merritt Island impoundments is complex and dynamic. Freshwater areas often feature submerged aquatic vegetation such as Yellow Water-Lily, Coontail, Water Hyacinth, and Eelgrass, which provide food and cover for fish and invertebrates. These, in turn, support higher trophic levels, including American Alligators, River Otters, Bobcats, and various birds. The more brackish impoundments may host salt-tolerant plants such as Red, Black, and White Mangroves, Black Needlerush, and Widgeon Grass. The brackish

The Anhinga (*Anhinga anhinga*), also known as the snakebird or water turkey, is a distinctive and common waterbird found along Florida's coastline. Anhingas are close relatives of cormorants. With its long, S-shaped neck and daggerlike bill, the Anhinga is an expert fisher. As with familiar cormorants such as the Double-Crested Cormorant, the Anhinga's feathers are not waterproof, which allows the birds to make sustained dives without the need to fight buoyancy. Both Anhingas and cormorants inhabit freshwater and saltwater environments, including swamps, mangroves, and coastal lagoons. This bird is a juvenile Anhinga.

A swamp forest of Sabal Palmettos, Live Oaks, and Red Maples rings a freshwater pond on Merritt Island. Saw Palmettos line the pond banks. The floating leaves are mostly Spatterdock (*Nuphar advena*), a water lily that produces a yellow flower when in bloom.

An isolated Live Oak–Cabbage Palm hammock in a freshwater marsh along Merritt Island's famous Black Point Wildlife Drive. The marshes in this area are mostly composed of Black Needlerush and various sedges, along with isolated Giant Leather Ferns, as seen at the bottom right corner here.

The Great Egret (*Ardea alba*) breeds in colonies along the Southeastern coast, typically nesting in trees near or over water. During breeding season, both male and female Great Egrets develop long, filamentous plumes called aigrettes. These delicate feathers play a role in courtship displays, as with the bird shown here in full display. During the breeding season, the bird's normally yellow lores (the bare area in front of the eye) turn a bright lime green.

and salty impoundments also support fish species that can adapt to changing salinities, such as the Red and Black Drums, Crevalle Jack, immature Tarpon, and Spotted Seatrout.

These impoundments undergo significant changes seasonally. Merritt Island Refuge managers often manipulate water levels and salinities to create ideal conditions for different species throughout the year. For instance, drawdowns can concentrate prey for wading birds and expose mudflats for shorebirds, while higher water levels benefit migratory and wintering diving ducks and fish populations.

The refuge also encompasses vast swaths of hardwood hammocks (islands of trees isolated in the marshes) and pine flatwoods, adding to its biological diversity. The oak palm hammocks are often barely above water level and are a kind of swamp habitat (*see illustration, p. 249*).

Black Needlerush
Juncus gerardii

Freshwater impoundments at Merritt Island

The Merritt Island National Wildlife Refuge features complex freshwater impoundments supporting various plant and animal species. These managed wetlands are home to numerous aquatic plants, including Spatterdock and Yellow Water-Lily, and multiple species of rushes and

American Alligator
Alligator mississippiensis

The Roseate Spoonbill (*Platalea ajaja*) is a striking wading bird found along the Atlantic Coast from South Carolina south to the Florida Keys. Known for its vibrant pink plumage and distinctive spoon-shaped bill, the Spoonbill forages in shallow waters for small fish and crustaceans. These birds are often seen in flocks, creating a beautiful spectacle in mangrove swamps and both freshwater and saltwater marshes. Roseate Spoonbills were once threatened by hunting and habitat loss, but conservation efforts have helped their populations recover.

sedges such as Black Needlerush that thrive in shallow waters. Along the elevated edges of the impoundments, you can find stands of Red Maple, Sabal Palmetto, Slash Pine, and Sweetgum trees. Saw Palmetto is found in drier areas.

The impoundments provide crucial habitat for many bird species. Wading birds such as Great Blue Herons, Snowy Egrets, and White Ibises frequent these areas to feed on small fish and amphibians. During migration seasons, the refuge becomes a haven for waterfowl including Blue-Winged Teal, Northern Pintail, and American Wigeon. The endangered Wood Stork also relies on these managed wetlands for nesting and foraging.

Reptiles are well represented in the impoundments, with American Alligators a common sight. Various turtle species inhabit these waters, including the Peninsula Cooter and Florida Softshell Turtle. The wetlands also support amphibians such as the Green Treefrog and Southern Leopard Frog.

The impoundments' freshwater fish populations include Largemouth Bass, Bluegill, and schools of the tiny white Mosquitofish. These fish serve as a food source for many of the ecosystem's birds, reptiles, and mammals, such as the River Otters that hunt in the waters.

The water management strategy at Merritt Island National Wildlife Refuge is designed to mimic natural tidal and hydrological cycles while also addressing specific conservation needs of migratory birds

Spatterdock
Nuphar lutea

Sabal Palmetto
Sabal palmetto

Large flocks of White Pelicans
(*Pelecanus erythrorhynchos*) are a common sight in the cooler months at Merritt Island NWR, where they winter in both freshwater and saltwater marshes.

Herons and Egrets of the Atlantic Coastline

Herons and egrets play crucial ecological roles in both the freshwater and brackish wetland environments along the Atlantic Coast. Common species such as the Great Blue Heron, Snowy Egret, Green Heron, and Tricolored Heron are vital components of coastal and inland wetland communities. These wading birds serve as top predators in wetlands, helping to regulate fish, crustacean, and insect populations.

Great Blue Heron
Ardea herodias

See p. 79 for the
Black-Crowned Night-Heron
Yellow-Crowned Night-Heron

Green Heron
Butorides virescens

Little Blue Heron (adult)
Egretta caerulea

Snowy Egret
Egretta thula

Great Egret
Ardea alba

Cattle Egret
Bubulcus ibis

Tricolored Heron
Egretta tricolor

Southern Leopard Frog
Lithobates sphenocephalus

and other animals that use the refuge. Refuge managers use a system of pumps, canals, and sluice gates to manipulate water levels in the impoundments throughout the year. This allows the managers to create optimal conditions for target species at different times, such as lowering water levels to concentrate prey for wading birds or maintaining higher levels to support submerged aquatic vegetation and fish populations.

Oak palm hammocks

The oak palm hammocks of Merritt Island NWR represent a unique and ecologically significant ecosystem. These distinctive, almost junglelike habitats are elevated areas within the coastal landscape that rise above the surrounding marshes and wetlands. The elevation of the hammocks varies from almost swamplike wetland terrain to drier areas with trails you can walk through. The word "hammock" probably originated as a sailor's term for a forested island amid marshes.

The oak palm hammocks owe their unique plant and animal communities to this slightly higher elevation. The elevation, often just a few feet above sea level, provides a crucial reprieve from the frequent

Boat-Tailed Grackles (*Quiscalus major*) are a common sight along Florida's coasts and wetlands, their raucous calls echoing across marshes and beaches. These large, glossy black birds gather in noisy flocks, their long keel-shaped tails distinguishing them from other grackle species. Highly adaptable, Boat-Tails thrive in both natural habitats and urban areas. Their gregarious nature and constant chatter make them a lively presence in Florida's coastal ecosystems.

A path through one of the drier oak palm hammocks at Merritt Island National Wildlife Refuge. A junglelike atmosphere prevails under a canopy of Live Oaks, Red Maples, Sweetgums, and Sabal Palmettos. These dense hammock islands and wetlands are ideal refuges for all kinds of wildlife—all but a few are isolated in the marshes and practically inaccessible to humans.

Live Oak leaves
Quercus virginiana

Live Oak trees

Great Blue Heron
Ardea herodias

flooding experienced in the surrounding wetlands. The soil in these hammocks, typically sandy but enriched with organic matter from years of leaf litter decomposition, creates a fertile substrate for a diverse plant life.

The structure of the hammocks is defined by a complex, multilayered canopy. The uppermost layer is dominated by towering Live Oaks and Sabal Palmettos, with the presence of Slash Pines indicating higher ground in Southeastern marshlands. These hammock trees form a dense canopy reaching 60 feet or more, creating a shaded understory below. The Live Oaks, with their sprawling branches draped in Spanish Moss, play a vital role in the ecosystem, providing shelter and food for various wildlife.

Beneath the main canopy, a subcanopy layer consists of smaller trees and tall shrubs. Common species in the subcanopy include Redbay, Wild Olive, Yaupon Holly, and American Beautyberry. These midstory plants contribute to the hammock's structural complexity, offering additional wildlife niches and helping regulate the forest's microclimate.

The understory of the oak palm hammocks is often less dense than that of other forest types due to the heavy shade cast by the upper layers. However, the hammocks support a variety of shade-tolerant ground and understory plants. Ferns are abundant, such as the Sword

Fern and Resurrection Fern, which often grows on the branches of live oaks; the Netted Chainfern thrives in humid, shaded conditions. The Giant Leather Fern appears frequently, along the edges of hammocks that get more sunlight. Saw Palmetto is probably the most common understory plant in both swampy and dry hammocks and can form dense thickets under the tree canopy, along with dense vines including Muscadine Grape, Virginia Creeper, and Poison Ivy.

Yaupon Holly
Ilex vomitoria

American Beautyberry
Callicarpa americana

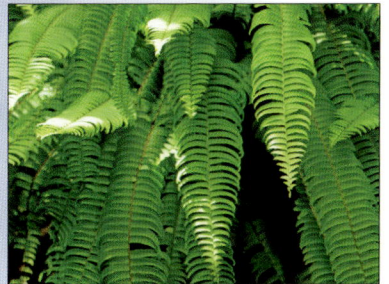

Florida (Native) Sword Fern
Nephrolepis exaltata

White Ibis
Eudocimus albus

Giant Leather Fern
Acrostichum danaeifolium

A vibrant freshwater pond at the Merritt Island NWR Visitor Center, where an excellent boardwalk trail leads visitors through a Southern Wild Rice (*Zizania aquatica*) marsh and into a Live Oak–Cabbage Palm–Red Maple swamp very similar to the wetter hammock communities at the refuge. These ponds and swamps play a critical role in flood control, both at the refuge and at the nearby Kennedy Space Center.

Common trees and shrubs in Southeast coastal freshwater wetland areas.

BALD CYPRESS *Taxodium distichum*

LIVE OAK *Quercus virginiana*

LIVE OAK LEAVES

SABAL PALMETTO *Sabal palmetto*

SAW PALMETTO *Serenoa repens*

RED MAPLE *Acer rubrum*

SLASH PINE *Pinus elliottii*

DAHOON HOLLY *Ilex cassine*

LOBLOLLY BAY *Gordonia lasianthus*

n invasive species

BRAZILIAN PEPPERTREE *S. terebinthifolia*

SWAMP DOGWOOD *Cornus foemina*

ELDERBERRY *Sambucus canadensis*

BEAUTYBERRY *Callicarpa americana*

BUTTONWOOD *Conocarpus erectus*

SWAMP TUPELO *Nyssa biflora*

SPANISH MOSS on a Live Oak

RESURRECTION FERN *P. polypodioides*

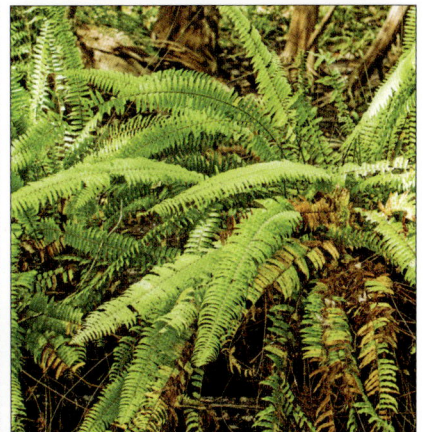

GIANT SWORD FERN *N. biserrata*

Common plants in Southeast coastal freshwater wetland areas.

SAWGRASS *Cladium jamaicense*

PICKERELWEED *Pontederia cordata*

MAIDENCANE *Panicum hemitomon*

COMMON CATTAIL *Typha latifolia*

BUTTONBUSH *Cephalanthus occidentalis*

ARROWHEAD *Sagittaria lancifolia*

SOFT RUSH *Juncus effusus*

SAND CORDGRASS *Spartina bakeri*

BLADDERWORT *Utricularia sp.*

WHITE WATER-LILY *Nymphaea alba*

SMARTWEED *Polygonum sp.*

BULRUSH *Scirpus sp.*

DUCKWEED *Lemna minor*

SPATTERDOCK *Nuphar advena*

FIRE FLAG *Thalia geniculata*

WAX MYRTLE *Myrica cerifera*

REDBAY *Persea borbonia*

MUSCADINE GRAPE *Vitis rotundifolia*

Bissell Road Duck Pond Trail, White Memorial Conservation Center, Litchfield, Connecticut

FURTHER READING

Not a long bibliography, but a few chosen books I've found to be particularly useful.

General information on wetlands and coastal regions

Alden, P., and B. Cassie. 1998. *National Audubon Society Field Guide to New England*. New York: Knopf.

Caduto, M. J. 1990. *Pond and Brook: A Guide to Nature in Freshwater Environments*. Hanover, NH: University Press of New England.

DeGraaf, R. M., and M. Yamasaki. 2001. *New England Wildlife: Habitat, Natural History, and Distribution*. Hanover, NH: University Press of New England.

Dodds, W., and M. Whiles. 2010. *Freshwater Ecology: Concepts and Environmental Applications of Limnology*. 2nd ed. New York: Elsevier.

Frankenberg, D. 1995. *The Nature of the Outer Banks: Environmental Processes, Field Sites, and Development Issues, Corolla to Ocracoke*. 2nd ed. Chapel Hill: University of North Carolina Press.

Hammerson, G. A. 2004. *Connecticut Wildlife: Biodiversity, Natural History, and Conservation*. Lebanon, NH: University Press of New England.

Kaufman, K., and K. Kaufman. 2012. *Kaufman Field Guide to Nature of New England*. Boston: Houghton Mifflin.

Mitsch, W., and J. G. Gosselink. 2015. *Wetlands*. Hoboken, NJ: Wiley.

Mitsch, W., J. G. Gosselink, C. J. Anderson, and Li Zhang. 2009. *Wetland Ecosystems*. Hoboken, NJ: Wiley.

Moss, B. 2017. *Ponds and Small Lakes*. Exeter, UK: Pelagic.

Myers, R., and J. Ewel. 1990. *Ecosystems of Florida*. Orlando: University of Central Florida Press.

Niering, W. A., and R. H. Goodwin. 1973. *Inland Wetland Plants of Connecticut*. New London: Connecticut Arboretum.

Schwarzman, B. 2002. *The Nature of Cape Cod*. Hanover, NH: University Press of New England.

Sperduto, D., and B. Kimball. 2011. *The Nature of New Hampshire: Natural Communities of the Granite State*. Hanover, NH: University Press of New England.

Thompson, E., E. Sorenson, and R. Zaino. 2019. *Wetland, Woodland, Wildland: A Guide to the Natural Communities of Vermont*. White River Junction, VT: Chelsea Green.

Travis, T. and S. Brown. 2014. *Pocketguide to Eastern Wetlands*. Mechanicsburg, PA: Stackpole.

Ponds and lakes

Jacobs, R., and E. O'Donnell, 2012. *A Fisheries Guide to the Lakes and Ponds of Connecticut.* Hartford: Connecticut Department of Energy and Environmental Protection.

Reid, G. K., H. R. Zim, and G. S. Fichter. 1967. *Pond Life: A Guide to Common Plants and Animals of North American Ponds and Lakes.* New York: St. Martin's.

Marshes and swamps

Eastman, J. 1992. *The Book of Forest and Thicket: Trees, Shrubs, and Wildflowers of Eastern North America.* Mechanicsburg, PA: Stackpole.

Magee, D. W. 1981. *Freshwater Wetlands: A Guide to Common Indicator Plants of the Northeast.* Amherst: University of Massachusetts Press.

Padgett, D. 2016. *Wetland Plants of New England: A Guide to Trees, Shrubs, and Lianas with Summer and Winter Keys.* Middleboro, MA: Spatterdock.

Travis, T., and S. Brown, 2014. *Pocket Guide to Eastern Wetlands.* Mechanicsburg, PA: Stackpole.

Vernal pools

Colburn, E. 2004. *Vernal Pools: Natural History and Conservation.* Newark, OH: McDonald and Woodward.

Johnson, S. D. 2021. *Vernal Pools: Documenting Life in Temporary Ponds.* North American Nature Photography Association. https://nanpa.org/resources/free-handbooks/.

Bogs

Davis, R. 2016. *Bogs and Fens: A Guide to the Peatland Plants of the Northeastern United States and Adjacent Canada.* Hanover, NH: University Press of New England.

Birds

Dunn, J., and J. Alderfer. 2017. *National Geographic Field Guide to the Birds of North America.* 7th ed. Washington, DC: National Geographic.

Gallo, F., 2018. *Birding in Connecticut.* Middletown, CT: Wesleyan University Press.

Peterson, R. T. 2010. *Peterson Field Guide to Birds of Eastern and Central North America.* 6th ed. Boston: Houghton Mifflin.

Sibley, D. 2017. *The Sibley Field Guide to Birds of Eastern North America.* 2nd ed. New York: Knopf.

Fish

Jacobs, R., and E. O'Donnell. 2009. *A Pictorial Guide to Freshwater Fish of Connecticut.* Hartford: Connecticut Department of Energy and Environmental Protection.

Page, L., and B. Burr. 2011. *Peterson Field Guide to the Freshwater Fishes of North America North of Mexico.* New York: Houghton Mifflin.

Insects

Dunkle, S. W. 2000. *Dragonflies Through Binoculars: A Field Guide to Dragonflies of North America.* New York: Oxford University Press.

Evans, A. V. 2008. *A Field Guide to Insects and Spiders of North America.* New York: Sterling.

Glassberg, J. 1999. *Butterflies Through Binoculars: The East.* New York: Oxford University Press.

Glassberg, J. 2017. *A Swift Guide to Butterflies of North America.* 2nd ed. Princeton, NJ: Princeton University Press.

O'Donnell, J., L. Gall, and D. Wagner. 2007. *The Connecticut Butterfly Atlas.* Hartford, CT: State Geological and Natural History Survey.

Amphibians

Powell, R., R. Conant, and J. Collins. 1991. *Peterson Field Guide to Reptiles and Amphibians of Eastern and Central North America.* Boston: Houghton Mifflin.

Raithel, C. 2019. *Amphibians of Rhode Island: Their Status and Conservation.* West Kingston: Rhode Island Division of Fish and Wildlife.

Mammals

Kays, R. W., and D. E. Wilson. 2009. *Mammals of North America.* 2nd ed. Princeton, NJ: Princeton University Press.

Wildflowers and wetland plants

Cox, D. 2002. *A Naturalist's Guide to Wetland Plants: An Ecology for Eastern North America.* Syracuse, NY: Syracuse University Press.

Elliman, T., and N. 2016. *Wildflowers of New England.* Portland, OR: Timber Press.

Newcomb, L., and G. Morrison. 1989. *Newcomb's Wildflower Guide.* Boston: Little, Brown.

Peterson, R. T., and M. McKenny. 1968. *A Field Guide to the Wildflowers of Northeastern and North-Central North America.* Boston: Houghton Mifflin.

Other books by this author

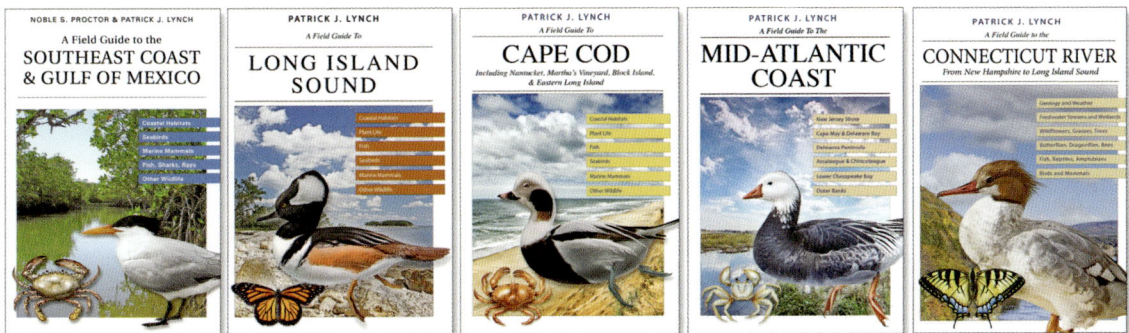

ILLUSTRATION CREDITS

All photography, artwork, diagrams, and maps are by the author, unless otherwise noted in this listing.

Additional photography credits

Images used with permission. All images are copyright 2026, by each source listed here. All rights reserved by the individual photographers and stock photography agencies.

Images used with permission from Twan Leenders

114–15 Common Frogs and Toad Species, 116–17 Common Salamander Species, 118–19 Common Turtle Species, 120–22 Common Snake Species, 186–87 Wood Frog Tadpole, 196 Bullfrog.

Images used under Creative Commons license from Judy Gallagher

39 Beaver, 64 Fragile Forktail, 133 Pied-Billed Grebe, 153 Painted Turtle, 177 Red Eft, 186–87 Wood Frogs, 186–87 Wood Frog Tadpole, 208 Five-Lined Skink, 237 Gopher Tortoise.

Images used under Creative Commons or Public Domain from Wikimedia Commons

2 Sundew, Agnieszka Kwiecień, Nova, 12 Flooding, Vermont National Guard, 59 Whirligig Beetles, James St. John, 60 Dragonfly Nymphs, Gail Hampshire, 64 Eastern Forktail, Rob Routledge, 65 Elfin Skimmer, Rob Routledge, 68 Hurricane Helene Damage, North Carolina, NCDOT, 69 Hurricane Helene Damage, North Carolina, NCDOT, 91 Woolgrass, Fredlyfish4, 95 Elodea, Christian Fischer, 154 Maidencane, USDA, 154 Swamp Lily, Vincent Lucas, 172 Eastern White Cedar, James St. John, 175 Two-Toed Ampiuma, Brian Gratwicke, 188 Phantom Midge Larva, Piet Spaans, 189 Fingernail Clams, Francisco Welter-Schultes, 189 Red Eft, Tom Walsh, 192 Balsam Fir, Robert H. Mohlenbrock, 192 Bog Willow, Rob Routledge, 192 Round-leaved Sundew, Agnieszka Kwiecień, 196 Leatherleaf, Krzysztof Ziarnek, 200 Bog Willow, Rob Routledge, 200 Leatherleaf, Krzysztof Ziarnek, 201 Creeping Snowberry, Superior National Forest, 201 Sundew, Agnieszka Kwiecień, 202 Swamp Loosetrife, Robert H. Mohlenbrock, 202 Rose Pagonia, wackybadger, 203 Pond Pine, autrpy, 208 Pond Pine, autrpy, 215 Cape Cod satellite photo, NASA LandSat Photo, 256 Maidencane, USDA.

Images used under license from Alamy

64 Spatterdock Darner, Clarence Holmes, 70 Green Frog Deformed, Michael Doolittle, 73 Flooded Trailer Park, Sipa USA, 142 Painted Turtle, Water-Frame, 142 Pond Split View, Marco Faggi, 142 Bullfrog, Franklin Kappa, 142 Tadpole, Herbert Frei, 202 Yellow Pitcher Plant, Botany Vision, 208 Yellow Pitcher Plant, Botany Vision, 220 Alewives, Milton Coghell, 221 Gulls and Alewives, Nature Photo Library, 257 Smartweed, Wirestock.

Images used under license from Adobe Stock

14 Volvox, micro_photo, 20 Eutrophic Pond, fotosenukas, 28 Particulate Organic Matter, Phoebe, 32 Daphnia, micro_photo, 42 Crayfish, ksena32, 47 Giant Water Bug, evergenesis, 47 Habitat Underwater, ead72, 47 Bluegills, RLS Photo, 47 Snapping Turtle, Christopher Seufert, 47 Watermilfoil, RLS Photo, 47 Spirogyra, Marius Burcea, 47 Bullfrog, sdbower, 48 River Otter, Stepan Jezek, 48 River Otter, Kletr, 48 Largemouth Bass, Phil Lowe, 48 Pike, prochym, 50 Canvasback Flock, flownalasla, 51 American Bullfrog Tadpole, Michael, 51 Backswimmer, float, 52 Water Lilies Underwater, Rostislav, 53 Fish X-ray, Leli, 53 Largemouth Bass, Lull, 54 Water Boatman, GRFischerjpeg, 54 Backswimmer, Eric Isselée, 56 Giant Water Bug, evergenesis, 56 Fishing Spider, saccobent, 56 Minnows, Duck Stock, 56 Largemouth Bass, Derrick, 57 Swallow, Jeff Huth, 57 Water Molecule, artegorov3@gmail, 57 Water Molecule, Imageflow, 58 Water Strider, janmiko, 60 Dragonfly nymph, Vitalli Hulai, 61 Dragonfly Laying Eggs, David, 62 Eastern Pondhawk, Anatolii, 62 Dragonfly Mating, Scott, 62 Dragonfly Nymph, Gonzalo, 62 Dragonfly Nymph Predation, Ernie Cooper, 62 Dragonfly Instar, Geza Farkas, 63 Teneral Stage, David, 63 Late Teneral Stage, easyparadise, 63 Dragonfly Nymph, Vitalii Hulai, 64 Emerald Spreadwing, Henk, 65 Azure Bluet, Doug Lemke, 65 Common Green Darner, Ivan Kuzman, 67 Rain Garden, Manyapha, 71 Water Pollution, boophuket, 75 Bullfrog Jumping, Ivan Kuzman, 81 Porro Binoculars, simmitorok, 84 Compass and Map, Stripped Pixel, 90 Arrowhead, Sandra Burm, 90 Common Cattail, dule964, 90 Narrow-Leaved Cattail, noppharat, 91 American Bur-Reed, nickkurzenko, 91 Ostrich Fern, scphoto48, 91 Royal Fern, knelson20, 91 Marsh Fern, Samuel, 91 Lady Fern, 92 Bloodroot, Margaret Burlingham, 92 Oxeye Daisy, bluehand, 92 Boneset, Erik, 92 Jack in the Pulpit, Dave, 92 Trout Lily, Gerry, 92 Red Trillium, Hamiza Bakirci, 92 Painted Trillium, Jack, 93 Cardinal Flower, nickkurzenko, 93 Pickerelweed, tamu, 93 White Water-Lily, Vitaliy, 93 Yellow Water-Lily, danilag, 94 Orange Hawkweed, skymoon13, 94 Ragged Robin, Imladris, 94 Fireweed, andreusK, 94 Swamp Rose, skymoon13, 94 New England Aster, Paul, 95 Coontail, illonajall, 95 Watermilfoil, linjerry, 95 Spirogyra, Marius Burcea, 95 Duckweed, Socoxbreed, 97 Redosier Dogwood, Carmen Hauser, 98 Paper Birch, Zimmerman, 98 Yellow Birch, David Katz, 98 American Hornbeam, Jon Benedictus, 98 Black Ash, Diana Samson, 98 Pussy Willow, Bluebird, 99 Sycamore, Batya, 99 Silver Maple, Aliaksandr Kisel, 99 Boxelder Maple, Anghi, 100 Water Strider, janmiko, 100 Whirligig Beetles, Olvita, 100 Backswimmer, float, 100 Water Boatman, GRFischer, 100 Giant Water Bug, evergenesis, 100 Diving Beetle, phototrip, 100 Fishing Spider, Michael, 100 Crayfish, AnikS, 100 Dragonfly Larva, Vitalii Hulai, 100 Freshwater Mussel, AB Photography, 100 Mayflies, gordzam, 101 Ebony Jewelwing, J. Stone, 101 Familiar Bluet, Samuel, 101 Northern Spreadwing, phototrip, 101 Eastern Forktail, Ray Akey, 101 Eastern Red Damsel, fabiosa_93, 101 River Jewelwing, rbkelle, 108 Acadian Hairstreak, Beth Baisch, 108 Banded Hairstreak, Mark, 108 Coral Hairstreak, Brian Lasenby, 108 Gray Hairstreak, Annette Shaff, 108 Juniper Hairstreak, Paul Sparks, 108 Eastern Tailed-Blue, Randy Anderson, 108 Little Wood-Satyr, Paul Sparks, 108 Common Ringlet, Riverwalker, 108 American Copper, Andrea Izzotti, 123 Black-Crowned Night-Heron, rpferreira, 123 Bittern, FotoRequest, 123 Common Loon, Brian Lasenby, 123 Double-Crested Cormorant, elharo, 124 Red-Shouldered Hawk, William, 124 Bald Eagle, Glenn, 124 Great Blue Heron, Jeffrey, 124

Northern Harrier, Fritz, 125 Barred Owl, Ron Dubreuil, 125 Killdeer, Brian E. Kushner, 125 Wilson's Snipe, amajk, 125 Woodcock, Steve Byland, 126 Mallard, havana1234, 126 Pintail, shaftinaction, 126 Sora, raptorcaptor, 126 Virginia Rail, Michael W. Potter, 127 Common Merganser, andreanita, 127 Black Duck, dule964, 127 Snow Goose, Tony Campbell, 128 Northern Waterthrush, Alex Papp, 128 Swamp Sparrow, Chase D'Animulls, 128 Carolina Wren, Brian E. Kushner, 128 White-Throated Sparrow, tmtracey720, 129 Prothonotary Warbler, Ray Hennessey, 129 Black-Capped Chickadee, Glen, 130 American Mink, serhio777, 130 Muskrat Swimming, Юрий Балагула, 130 Beaver swimming, Christian Musat, 130 Porcupine, hkuchera, 130 Black Bear, rima15, 131 River Otter Swimming, patrick, 131 Raccoon Swimming, AB Photography, 131 Muskrat, , 138–39 Wood Duck, Ivan Kuzmin, 138–39 Pickerelweed, Sermek, 138–39 Muskrat, byrdyak, 138–39 Virginia Rail, Michael W. Potter, 140 Virginia Rail, 489761197, 140 Bittern, Paul, 140 Sora, raptorcaptor, 142–43 Green Heron, SunnyS, 142–43 Dragonfly, Chase D'Animulls, 142–43 Northern Pike, prochym, 142–43 Green Darner, , 144 Beaver-Chewed Tree, Kaloa, 144–45 Brown Rat, Eric Issele_e, 144–45 Muskrat, Iri-sha, 144–45 Raccoon, rabbit75, 144–45 River Otter, Cloudtail, 144–45 Beaver, jnjhuz, 148 Beaver on Dam, Ernie Howard, 152 Cattails, schapinskaja, 152 Narrow-Leaved Cattails, nopharat, 152 Bulrush, Katarzyna, 152 White Water-Lily, Vitally Hrabar, 152 Yellow Water-Lily, danilag, 152 Common Duckweed, Socoxbreed, 152 Arrowhead, Sandra Burm, 153 Green Frog, Steve Byland, 153 Muskrat, Tspider, 153 Beaver, Frank Fichtmuller, 154 Sawgrass, Robert Miller, 154 Common Cattail, dule964, 154 Alligator Flag, kendonice, 154 Arrowhead, Sandra Burm, 154 Water Hyacinth, Maule, 155 Florida Gar, M-Production, 155 Mosquitofish, Stan, 155 Green Treefrog, donyanedonam, 157 Common Yellowthroat, Agami, 160 Redosier Dogwood, Carmen Hauser, 167 Bald Cypress Swamp, Danita Delimont, 172 Balsam Fir, simona, 172 Swamp Azalea, lorenza62, 172 Atlantic White Cedar, 173 Eastern Newt, Hamilton, 173 Spotted Salamander, ondreika, 173 Northern Watersnake, ondreika, 173 Wood Duck, Ivan Kuzmin, 173 Swamp Sparrow, Chase D'Animulls, 173 Muskrat, Mircea Costina, 174 Bald Cypress Swamp, Danita Delimont, 174 Water Tupelo, Gerry, 174 Sweet Gum, simona, 174 Swamp Chestnut, Jon Benedictus, 174 Willow Oak, Alexander Denisenko, 174 Swamp Dogwood, Garmon, 174 Swamp Azalea, lorenza62, 175 Green Treefrog, donyanedonam, 175 Pig Frog, Joy, 175 Cottonmouth, Kris, 175 Brown Water Snake, LP, 175 Wood Duck, Ivan Kuzmin, 175 Prothonotary Warbler, Bob, 175 Swamp Rabbit, Ivan Kuzmin, 180 Eastern Newt, Hamilton, 181 Black-Crowned Night-Heron, Robert, 181 Fairy Shrimp, phototrip.cz, 181 Dragonfly Larva, bennytrapp, 181 Spring Peeper, ondreika, 182–83 Eastern Comma, pimmimemom, 182–83 Wood Frog, ondreika, 182–83 Wood Frog Egg Mass, Mark Lotterhand, 182–83 American Toad, ondreika, 182–83 Spring Peeper, ondreika, 182–83 Gray Treefrog, Natalia Kuzmina, 182–83 Spotted Salamander, ondreika, 182–83 Red-Spotted Newt, Melinda Fawver, 182–83 Common Garter Snake, Nynke, 182–83 Snapping Turtle, Mark Lotterhand, 186–87 Wood Frog, ondreika, 186–87 Wood Frog Egg Mass, Mark Lotterhand, 188 Water Strider, janmiko, 188 Whirligig Beetles, Olvita, 188 Backswimmer, float, 188 Giant Water Bug, evergenesis, 188 Fairy Shrimp, phototrip.cz, 188 Water Boatman, GRFischerjpeg, 188 Caddisfly Larva, dule964, 188 Diving Beetle, phototrip, 189 Eastern Newt, Hamilton, 189 Jefferson Salamander, Tom Wilhelm, 189 Green Frog, Steve Byland, 189 American Toad, ondreika, 189 Spring Peeper, Wirepec, 189 Gray Treefrog, Natalia Kuzmina, 190 Purple Pitcher Plant, Nina, 191 Sundew, Matauw, 192–93 Pitcher Plant, martinrossi, 192–93 Grass Pink, scandamerican, 192–93 Labrador Tea, Amelia, 194 Purple Pitcher Plant, anjahennern, 196–97 Bog Turtle, ondreika, 200 Balsam Fir, simona, 200 American Larch, steadb, 200 Red Spruce, Capucine Dieppedale, 200 Northern White Cedar, 201 Labrador Tea, Amelia, 201 Sweetgale, BestPhotoStudio, 201 Cranberry, pisotckii, 201 Bog Rosemary, lembrechtsjonas, 201 Royal Fern, Knelson20, 201 Spoonleaf Sundew, brudertack69, 202 Starflower, Garry, 202 Pitcher Plant, martinrossi, 202 Grass Pink, scandamerican, 202 Yellow Pitcher Plant Flowers, Iwona, 202 Venus Flytrap, russell102, 203 Pond Pine Cones, Cuteldeas, 203 Prothonotary Warbler, Samuel, 203 Red-Cockaded Woodpecker, beng, 203 Raccoon, redav, 203 Black Bear, Georgia, 203 Red Wolf, Abeselom Zerit, 204 Red Spruce, Capucine Dieppedale, 204 American Larch, steadcb, 208–9 Venus Flytrap, russell102, 208–9 Greentree Frog, donyanedonam, 211 Alligator River NWR, Moelyn Photos, 219 Walden Pond Shoreline, Alizada Studios, 237 Scrub Jay, Joseph, 245 Spatterdock, Robert, 251 Yaupon Holly, Janelle, 251 Sword Fern, Ika, 251 Giant Leather Fern, Jaimie Tuchman, 254 Bald Cypress, 90783759, 254 Live Oak, Melissa, 254 Saw Palmetto, Wirepec, 254 Slash Pine, Sunshower Shots, 254 Loblolly Bay, Donna Bollenbach, 255 Brazilian Peppertree, Wagner Campelo, 255 Swamp Dogwood, Garmon, 255 Elderberry, Volodymyr, 255 Buttonwood, Sunshower Shots, 255 Swamp Tupelo, Camille Lamoureux, 255 Resurrection Fern, Careth, 255 Giant Sword Fern, duke215, 256 Sawgrass, Robert Miller, 256 Common Cattail, schapinskaja, 256 Arrowhead, Sandra Burm, 256 Soft Rush, Vtigrapher, 256 Bladderwort, Oksana, 257 Bulrush, Katarzyna, 257 Duckweed, socoxbreed, 257 Spatterdock, Robert, 257 Fire Flag, kendonice.

Images used under license from Dollar Photo Club (now owned by Adobe)

43 Painted Turtles, 123 Belted Kingfisher, 124 Red-Tailed Hawk, 124 Osprey, 126 Wood Duck, 127 Hooded Merganser, 127 Lesser Scaup, 128 Yellow Warbler, 128 Song Sparrow, 128 Marsh Wren, 129 Common Grackle, 129 American Goldfinch, 129 American Robin, 130 Eastern Cottontail, 131 Red Squirrel, 139 Kestrel, 138 Screech Owl, 155 Snapping Turtle, 192 Round-Leaved Sundew.

Images used under license from Shutterstock

83 Lone Star Tick, ondreika, 83 Black-Legged Tick, Chris Ritchie Photo, 83 Dog Tick, Melinda Fawver, 100 Pond Leech, frank60, 126 Green-Winged Teal, Erni, 127 American Coot, Martha Marks, 128 Red-Winged Blackbird, BGSmith, 153 Red-Winged Blackbird, BGSmith, 189 Dragonfly Nymph, Kunfu01, 202 Dragon's Mouth, Lee Ellsworth.

Northern Leopard Frog (*Lithobates pipiens*), Hamden, Connecticut.

Index

Branford Supply Pond, Branford, Connecticut.